THE PHANEROZOIC CARBON CYCLE

To Jacques Joseph Ebelmen, who had it all figured out 160 years ago and whose pioneering work on the long-term carbon cycle is virtually unknown

THE PHANEROZOIC CARBON CYCLE: CO_2 AND O_2

Robert A. Berner

UNIVERSITY PRESS

2004

OXFORD
UNIVERSITY PRESS

Oxford   New York
Auckland   Bangkok   Buenos Aires   Cape Town   Chennai
Dar es Salaam   Delhi   Hong Kong   Istanbul   Karachi   Kolkata
Kuala Lumpur   Madrid   Melbourne   Mexico City   Mumbai   Nairobi
São Paulo   Shanghai   Taipei   Tokyo   Toronto

Published by Oxford University Press, Inc.
198 Madison Avenue, New York, New York 10016

www.oup.com

Oxford is a registered trademark of Oxford University Press

Library of Congress Cataloging-in-Publication Data
Berner, Robert A., 1935–
The phanerozoic carbon cycle : CO_2 and O_2 / Robert A. Berner.
p.   cm.
Includes bibliographical references and index.
ISBN 0-19-517333-3
1. Atmospheric carbon dioxide—Evolution.   2. Carbon cycle (Biogeochemistry)   I. Title.
QC879.8.B47 2004
577'.144—dc22   2003060954

9 8 7 6 5 4 3 2 1

Printed in the United States of America
on acid-free paper

Preface

There is much confusion attached to the term "carbon cycle." It has been applied to different time scales ranging from hours in biological systems, to decades in future global warming, to millennia and hundreds of millennia in climate history. Much neglected is the cycling of carbon over longer time scales, and the purpose of this book is to alleviate this situation. What I call the "long-term carbon cycle" involves the exchange of carbon between rocks and the various reservoirs near the earth's surface, the latter including the atmosphere, hydrosphere, biosphere, and soils. Exchange with the surface involves such processes as chemical weathering of silicate minerals, burial of organic matter in sediments, and volcanic degassing of CO_2. I have spent much time worrying about such processes and feel that it is time to show how the long-term cycle works and how to use it in deducing factors affecting the evolution of atmospheric CO_2 and O_2 over the past 550 million years (Phanerozoic time). This is a new world to most people studying the "carbon cycle," especially as it relates to future global warming. It is not generally realized that global warming due to the burning of fossil fuels is simply a large acceleration of one of the major processes of the long-term carbon cycle, the oxidative weathering of sedimentary organic matter.

Descriptive discussion of the long-term carbon cycle is not enough. The other role of this book is to show how one can make quantitative estimates of rates of carbon flux between rocks and the earth's surface and how these fluxes can be used to estimate past levels of atmospheric CO_2 and O_2. In this way, I introduce the reader to a much needed multidisciplinary quantitative approach to earth history, which is sometimes referred to as "earth system science." I, and other workers, have published a number of papers on modeling of the long-term cycle, but there is no central place one can go to get the fundamentals of this cycle. This book is hopefully that place.

I am indebted to the many discussions of the long-term cycle with earth scientists, which are too numerous to list here. However, discussions with Klaus Wallmann, David Beerling, Dana Royer, Tom Crowley, Steve Petsch, Derrill Kerrick, Ken Caldeira, Leo Hickey, Dick Holland, Bill Hay, Fred Mackenzie, Bette Otto-Bliesner, Betty Berner, John Hedges, John Hayes, Lee Kump, and Tony Lasaga at various times over the past 20 years have been unusually helpful. Several of these people will recognize their contribution to the GEOCARB modeling discussed in this book. Special acknowledgment goes to the late Bob Garrels, who introduced me to geochemical cycle modeling in general. Without his influence this book would never have been written. Also, the book would probably not have been written now if editor Cliff Mills, at the suggestion of Brian Skinner, hadn't suggested doing so.

Contents

THE PHANEROZOIC CARBON CYCLE

1

Introduction

The cycle of carbon is essential to the maintenance of life, to climate, and to the composition of the atmosphere and oceans. What is normally thought of as the "carbon cycle" is the transfer of carbon between the atmosphere, the oceans, and life. This is not the subject of interest of this book. To understand this apparently confusing statement, it is necessary to separate the carbon cycle into two cycles: the short-term cycle and the long-term cycle.

The Short-Term Carbon Cycle

The "carbon cycle," as most people understand it, is represented in figure 1.1. Carbon dioxide is taken up via photosynthesis by green plants on the continents or phytoplankton in the ocean. On land carbon is transferred to soils by the dropping of leaves, root growth, and respiration, the death of plants, and the development of soil biota. Land herbivores eat the plants, and carnivores eat the herbivores. In the oceans the phytoplankton are eaten by zooplankton that are in turn eaten by larger and larger organisms. The plants, plankton, and animals respire CO_2. Upon death the plants and animals are decomposed by microorganisms with the ultimate production of CO_2. Carbon dioxide is exchanged between

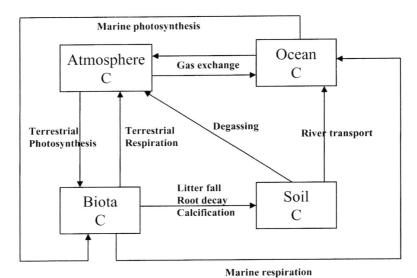

Figure 1.1. The short-term carbon cycle. (Adapted from Berner, 1999.)

the oceans and atmosphere, and dissolved organic matter is carried in solution by rivers from soils to the sea. This all constitutes the short-term carbon cycle. The word "short-term" is used because the characteristic times for transferring carbon between reservoirs range from days to tens of thousands of years. Because the earth is more than four billion years old, this is short on a geological time scale.

As the short-term cycle proceeds, concentrations of the two principal atmospheric gases, CO_2 and CH_4, can change as a result of perturbations of the cycle. Because these two are both greenhouse gases—in other words, they adsorb outgoing infrared radiation from the earth surface—changes in their concentrations can involve global warming and cooling over centuries and many millennia. Such changes have accompanied global climate change over the Quaternary period (past 2 million years), although other factors, such as variations in the receipt of solar radiation due to changes in characteristics of the earth's orbit, have also contributed to climate change. Over the past century human perturbation of the short-term carbon cycle, from activities such as deforestation and biomass burning (for CO_2), and rice cultivation and cattle raising (for CH_4), have contributed to a rise in atmospheric levels of these gases. However, the major perturbation of the level of atmospheric CO_2, and consequently an overall rise in global temperature over the past century, is due to a process of the long-term carbon cycle. This is the burning of fossil fuels. Organic matter in sedimentary rocks, which would otherwise be slowly exposed to the atmosphere by erosion and oxidized

by weathering, is instead being rapidly removed from the ground, oxidized by burning, and given off to the atmosphere as CO_2.

The Long-Term Carbon Cycle

Over millions of years carbon still undergoes constant cycling and recycling via the short-term cycle, but added to this is a new set of processes affecting carbon. This is the long-term carbon cycle, the subject
of this book. What distinguishes the long-term carbon cycle from the
short-term cycle is the transfer of carbon to and from rocks. This is illustrated in figure 1.2. Over millions of years carbon transfers to and
from rocks can result in changes in atmospheric CO_2 that cannot be attained via the short-term carbon cycle. This is because there is so much
more carbon in rocks than there is in the oceans, atmosphere, biosphere,
and soils combined (table 1.1). The maximum change in atmospheric
CO_2 that could be obtained, for example by burning all terrestrial life
and equilibrating the resulting CO_2 with the oceans, would be less than
a 25% increase from the present level (Berner, 1989). In contrast, changes
in the long-term carbon cycle have likely resulted in past increases in
atmospheric CO_2 to levels more than 10 times the present levels, resulting in intense global warming (Crowley and Berner, 2001).

Let us go for a tour through the long-term cycle. As one will see, various aspects of the short-term cycle are components of the long-term

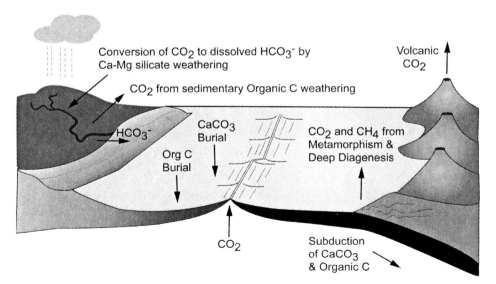

Figure 1.2. The long-term carbon cycle. (After Berner, 1999.)

Table 1.1. Masses of carbon involved in both the short-term (prehuman) and long-term carbon cycles compared with some fluxes in the long-term cycle.

Substance or flux	Mass (10^{18} mol)	Flux (10^{18} mol/my)
Carbonate C in rocks	5000	
Organic C in rocks	1250	
Oceanic dissolved inorganic carbon	2.8	
Soil carbon (including caliche)	0.3	
Atmospheric CO_2	0.06	
Terrestrial biosphere	0.05	
Marine biosphere	0.0005	
Organic C burial in sediments		5
CO_2 uptake by Ca and Mg silicate weathering		7
CO_2 release by volcanic degassing		3–9

Modified from Berner (1989, 1991).

cycle, but it is the participation of rocks that is critical. Atmospheric carbon dioxide is taken up by plant photosynthesis, and organic matter builds up in soils. Microbial decomposition in the soil leads to a buildup of organic acids and CO_2 in the soil. The organic acids and carbonic acid formed from CO_2 react with minerals in rocks to liberate cations and acid anions to solution, and the organic acid anions are oxidized to bicarbonate. Of special interest is reaction with calcium- and magnesium-containing silicate minerals. A representative overall reaction for a generalized calcium silicate is

$$2CO_2 + 3H_2O + CaSiO_3 \rightarrow Ca^{++} + 2HCO_3^- + H_4SiO_4 \qquad (1.1)$$

The dissolved species are carried by groundwater to rivers and by rivers to the sea. In the oceans the Ca^{++} and HCO_3^- are precipitated, mostly biogenically, as calcium carbonate:

$$Ca^{++} + 2HCO_3^- \rightarrow CaCO_3 + CO_2 + H_2O \qquad (1.2)$$

and the silicic acid as biogenic silica:

$$H_4SiO_4 \rightarrow SiO_2 + 2H_2O \qquad (1.3)$$

The calcium carbonate and biogenic silica are then buried in marine sediments and eventually into the geological record. Adding reactions (1.1), (1.2), and (1.3), we obtain the overall reaction:

$$CO_2 + CaSiO_3 \rightarrow CaCO_3 + SiO_2 \qquad (1.4)$$

This is a key reaction of the long-term carbon cycle and represents the transfer of carbon from the atmosphere to the rock record by means of weathering and marine carbonate sedimentation. The reaction was first deduced by Ebelmen (1845)[1] and much later by Urey (1952). It can just as well be written in terms of Mg and Ca-Mg silicates and carbonates. In this book the reaction will be referred to as the Ebelmen-Urey reaction. Only the weathering of Ca and Mg silicates is important; weathering of Na and K silicates does not lead to loss of CO_2 because these elements do not form common carbonate minerals in sediments. (The CO_2 consumed during Na and K silicate weathering is returned to the atmosphere during the formation of new Na and K silicates in sediments; see Mackenzie and Garrels, 1966). Also, weathering of Mg silicates does not necessitate the formation of Mg-containing carbonates. The dissolved Mg from silicate weathering, when delivered to the oceans, is well known to undergo a series of different reactions with submarine basalts that results in the liberation of Ca that is precipitated as $CaCO_3$ (Berner and Berner, 1996).

If reaction (1.4) were to continue alone, all atmospheric CO_2 would be removed in only about 10,000 years or, with resupply of CO_2 from the oceans, in about 300, 000 years (Sundquist, 1991). Over millions of years there must be a restoring process, and the principal one is the degassing of CO_2 to the atmosphere and oceans via the opposite of reaction (1.4). In other words, for our reference Ca silicate,

$$CaCO_3 + SiO_2 \rightarrow CO_2 + CaSiO_3 \qquad (1.5)$$

Reaction (1.5) represents decarbonation via volcanism, metamorphism, and diagenesis, and together reactions (1.4) and (1.5) and their magnesium silicate and carbonate analogues constitute the silicate-carbonate subcycle. This "reverse" reaction was also adduced by Ebelmen and Urey.

Reactions (1.1) to (1.5) are used to simplify representation of the silicate-carbonate subcycle. In reality weathering involves Ca and Mg aluminosilicates, such as calcic plagioclase, with aluminum precipitated as clay minerals. The clay minerals are then involved in reactions with calcium carbonate or dolomite to form igneous and metamorphic (and even diagenetic) silicates. But the overall principal of CO_2 uptake and realease is the same as represented by reactions (1.1)–(1.5).

So far the weathering of carbonates has not been mentioned. This is because, on a million-year time scale, it has little direct effect on atmospheric CO_2. This can be seen by the weathering reaction for calcium carbonate:

$$CO_2 + H_2O + CaCO_3 \rightarrow Ca^{++} + 2HCO_3^- \qquad (1.6)$$

---

1. J.J. Ebelmen, more than 100 years ahead of his time, deduced correctly almost all of the major long-term processes affecting atmospheric CO_2 and O_2, including volcanism, the role of plants in weathering, the weathering and burial of organic matter and pyrite, and the weathering of basalt (Berner and Maasch, 1996).

This is the reverse of reaction (1.2) for the precipitation of $CaCO_3$ in the oceans. Thus, the weathering of $CaCO_3$, followed by transport of Ca^{++} and HCO_3^- to the oceans and the precipitation of new $CaCO_3$, results in no net change in atmospheric CO_2. On shorter time scales (e.g., stages of the Pleistocene epoch), weathering of carbonates can be greater than, or less than, their precipitation from the oceans, with the excess carbon stored in or lost from seawater. However, over millions of years the necessary storage or loss becomes so excessive (the mean residence time for bicarbonate in the oceans is about 100,000 years; see Holland, 1978) that purely inorganic precipitation will occur or carbonate sediments cannot form. There is little evidence that such extreme conditions have ever occurred during the Phanerozoic which is marked by continuous deposition of limestones rich in biogenic skeletal debris (e.g., Stanley, 1999).

That the weathering of carbonates has no direct effect on atmospheric CO_2 does not mean that this process can be ignored in studying the long-term carbon cycle. This is because it is necessary to account for all sinks and sources of carbon, and carbonate weathering supplies carbon for transport from minerals to the oceans. (Note that in reaction 1.6 there are two bicarbonate ions produced from calcium carbonate weathering and that one of them comes from the carbon contained within the carbonate mineral itself.) Modeling of the long-term cycle involves calculation of the rate of Ca and Mg silicate weathering, and this requires a knowledge of the rates of Ca and Mg carbonate weathering (F_{wc} in equation 1.13 below).

The long-term carbon cycle has another component, the organic subcycle. This is represented by the reactions

$$CO_2 + H_2O \rightarrow CH_2O + O_2 \tag{1.7}$$

$$CH_2O + O_2 \rightarrow CO_2 + H_2O \tag{1.8}$$

Reaction (1.7) is normally thought to represent photosynthesis (short-term carbon cycle). In the long-term cycle it represents *net* photosynthesis (photosynthesis minus respiration) resulting in the burial of organic matter into sediments. It is the principal process of atmospheric O_2 production (Ebelmen, 1845). Reaction (1.8) represents, not respiration as normally understood, but "georespiration," the oxidation of old organic carbon in rocks. This georespiration occurs either by oxidative weathering of organic matter in shales and other sedimentary rocks uplifted onto the continents, or by the microbial or thermal decomposition of organic matter to reduced carbon containing gases, followed by oxidation of the gases upon emission to the atmosphere. An example of the latter is

$$2CH_2O \rightarrow CO_2 + CH_4 \tag{1.9a}$$

$$CH_4 + 2O_2 \rightarrow CO_2 + 2\,H_2O \tag{1.9b}$$

which together sum to reaction (1.8).

A special example of reaction (1.8) is the burning of fossil fuels by humans. Coal and oil are concentrated forms of sedimentary organic matter. Under natural processes the coal and oil is slowly oxidized by weathering and thermal degassing of hydrocarbons as mentioned above. However, humans have extracted these substances from the ground so quickly, from a geological perspective, that oxidation of the carbon occurs at a rate about 100 times faster than what would occur naturally. As a result the long-term carbon cycle impinges on the short-term cycle, and this has led to an extremely fast historic rise in atmospheric CO_2 (IPCC, 2001).

Modeling the Phanerozoic Carbon Cycle

Together the carbonate-silicate and organic long-term subcycles play the dominant role in controlling the levels of atmospheric CO_2 and O_2 over millions to billions of years. In this book I show how these subcycles have operated only over the past 550 million years, the Phanerozoic eon. The Phanerozoic is chosen because of the abundance of critical data such as abundant multicellular body fossils, relatively noncontroversial paleogeographic reconstructions, and relatively agreed-upon tectonic and climatic histories. Such a situation is not available for the Precambrian. The plethora of Phanerozoic geological, biological, and climatic data are extremely useful in trying to recreate the history of the carbon cycle. This will be done in the present book. The reader is referred to the books by Holland (1978, 1984) for discussion of the carbon cycle before the Phanerozoic.

All Phanerozoic carbon cycle models to date use analogous formulations for the mass balance of carbon added to and from the Phanerozic rock record (e.g., Budyko and Ronov, 1979; Walker et al., 1981; Berner et al, 1983; Garrels and Lerman, 1984; Berner, 1991, 1994; Kump and Arthur, 1997; Francois and Godderis, 1998; Tajika, 1998, Berner and Kothavala, 2001; Wallmann, 2001; Kashiwagi and Shikazono, 2003; Bergman et al., 2003; Mackenzie et al., 2003). The simplest approach to carbon mass balance modeling is to introduce the concept of the "surficial system" (Berner, 1994, 1999) consisting of the oceans + atmosphere + biosphere + soils (the reservoirs of the short-term cycle). A generalized mass balance expression for the surficial system is:

$$dM_c/dt = F_{wc} + F_{wg} + F_{mc} + F_{mg} - F_{bc} - F_{bg} \qquad (1.10)$$

where

M_c = mass of carbon in the surficial system

F_{wc} = carbon flux from weathering of Ca and Mg carbonates

F_{wg} = carbon flux from weathering of sedimentary organic matter

F_{mc} = degassing flux from volcanism, metamorphism, and diagenesis of carbonates

F_{mg} = degassing flux from volcanism, metamorphism and diagenesis of organic matter

F_{bc} = burial flux of carbonate-C in sediments

F_{bg} = burial flux of organic-C in sediments.

An additional mass balance expression for ^{13}C involving the stable isotopes of carbon has been found to be of great help in doing long-term carbon cycle modeling:

$$d(\delta_c M_c)/dt = \delta_{wc}F_{wc} + \delta_{wg}F_{wg} + \delta_{mc}F_{mc} \qquad (1.11)$$
$$+ \delta_{mg}F_{mg} - \delta_{bc}F_{bc} - \delta_{bg}F_{bg}$$

where $\delta = [(^{13}C/^{12}C) / (^{13}C/^{12}C)stnd - 1]$ 1000. and stnd represents a reference standard. Equations (1.10) and (1.11), when combined with assumptions about weathering, burial and degassing, can be used to calculate the various carbon fluxes as a function of time. More complicated expressions have been used for carbon mass balance in some models where the surficial system is broken up into its parts and separate mass balance expression are used for carbon in the atmosphere, biosphere, and ocean. However, the simpler approach of equations (1.10) and (1.11) will be emphasized in the present book. By lumping the atmosphere, oceans, life and soils together, processes involved in the short-term carbon cycle are avoided in the modeling, and the use of steady-state becomes possible. A diagrammatic presentation of this approach is shown in figure 1.3.

The weathering and degassing fluxes of carbon integrated over millions of years are much larger than the amount of carbon that can be stored in the surficial system (table 1.1). Adding excessive dissolved calcium and bicarbonate to the oceans eventually would result in the global inorganic precipitation of $CaCO_3$. (Adding too little calcium and bicarbonate would result eventually in an acid ocean and the inability to ever form limestones.) The area of land can hold just so much biomass and soil carbon. Too much CO_2 in the atmosphere leads to excessive warming due to the atmospheric greenhouse effect. Because of the inability to store much carbon in the surficial system, over millions of years one can assume that the carbon loss fluxes, due to organic carbon burial and Ca and Mg silicate weathering followed by Ca and Mg carbonate burial, are essentially balanced by degassing fluxes from thermal carbonate decomposition and organic matter oxidation (Berner, 1991, 1994; Tajika, 1998). In other words, there is a quasi steady state such that:

$$dM_c/dt = 0 \text{ and } d(\delta_c M_c)/dt = 0 \qquad (1.12)$$

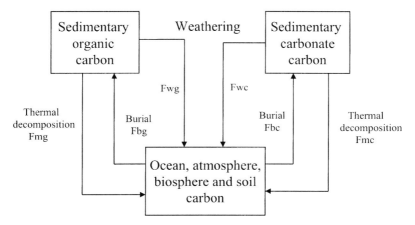

Figure 1.3. Modeling diagram for the long-term carbon cycle. F_{wc} = carbon flux from weathering of Ca and Mg carbonates; F_{wg} = carbon flux from weathering of sedimentary organic matter; F_{mc} = degassing flux from volcanism, metamorphism, and diagenesis of carbonates; F_{mg} = degassing flux from volcanism, metamorphism, and diagenesis of organic matter; F_{bc} = burial flux of carbonate-C in sediments; F_{bg} = burial flux or organic-C in sediments.

This greatly simplifies theoretical modeling of the long-term carbon cycle. It means that the sum of input fluxes to the surficial system are essentially equal to the sum of all output fluxes. For each million-year time step, although input and output fluxes of carbon to the surficial system may change, they quickly readjust during the time step to a new steady state, This is known as the quasistatic approximation. Non–steady-state modeling (Sundquist, 1991) has shown that perturbations from surficial system steady state, for the long-term carbon cycle, cannot persist for more than about 500,000 years.

At steady state, the CO_2 uptake flux to form HCO_3^- accompanying the weathering of Ca and Mg silicates F_{wsi} is determined from the mass balance expression for bicarbonate (reactions 1.1, 1.2, and 1.6):

$$F_{wsi} = F_{bc} - F_{wc} \qquad (1.13)$$

$F_{bc} - F_{wc}$ represents the carbonate that is formed only from the weathering of Ca and Mg silicates, as opposed to that formed from both Ca and Mg silicate and carbonate weathering (F_{bc}). Equation (1.13) illustrates the necessity of knowing the rate of carbonate weathering (F_{wc}) in calculating the rate of silicate weathering.

In GEOCARB (Berner, 1991, 1994; Berner and Kothavala, 2001) and similar modeling (e.g., Kump and Arthur, 1997; Tajika, 1998; Wallmann, 2001) the weathering and degassing fluxes, F_{wc}, F_{wg}, F_{mc}, F_{mg} are expanded in terms of nondimensional parameters representing how a variety of processes affect rates of weathering and degassing. The parameters are multiplied by present fluxes to obtain ancient fluxes. These non-dimensional parameters are discussed in the next three chapters and provide a window into the inner workings of the long-term Phanerozoic carbon cycle. The last two chapters show how calculations based on long-term carbon cycle modeling can be used to estimate the Phanerozoic evolution of atmospheric CO_2 and O_2. The modeling results are then compared to independent estimates of paleo-CO_2 and O_2 to give some idea of the accuracy and deficiencies of the modeling.

2

Processes of the Long-Term Carbon Cycle: Chemical Weathering of Silicates

Carbon dioxide is removed from the atmosphere during the weathering of both silicates and carbonates, but, over multimillion year time scales, as pointed out in chapter 1, only Ca and Mg silicate weathering has a direct effect on CO_2. Carbon is transferred from CO_2 to dissolved HCO_3^- and then to Ca and Mg carbonate minerals that are buried in sediments (reaction 1.4). In this chapter the factors that affect the rate of silicate weathering and how they could have changed over Phanerozoic time are discussed. Following classical studies (e.g., Jenny, 1941), the factors discussed include relief, climate (rainfall and temperature), vegetation, and lithology. However, over geological time scales, additional factors come into consideration that are necessarily ignored in studying modern weathering. These include the evolution of the sun and continental drift. The aim of this book is to consider all factors, whether occurring at present or manifested only over very long times, that affect weathering as it relates to the Phanerozoic carbon cycle.

Mountain Uplift, Physical Erosion, and Weathering

Within the past decade much attention has been paid to the effect of mountain uplift on chemical weathering and its effect on the uptake of atmospheric CO_2, an idea originally espoused by T.C. Chamberlin

(1899). The uplift of the Himalaya Mountains and resulting increased weathering has been cited as a principal cause of late Cenozoic cooling due to a drop in CO_2 (Raymo, 1991). Orogenic uplift generally results in the development of high relief. High relief results in steep slopes and enhanced erosion, and enhanced erosion results in the constant uncovering of primary minerals and their exposure to the atmosphere. In the absence of steep slopes, a thick mantle of clay weathering product can accumulate and serve to protect the underlying primary minerals against further weathering. An excellent example of this situation is the thick clay-rich soils of the Amazon lowlands where little silicate weathering occurs (Stallard and Edmond, 1983). In addition, the development of mountain chains often leads to increased orographic rainfall and, at higher elevations, increased erosion by glaciers. All these factors should lead to more rapid silicate weathering and faster uptake of atmospheric CO_2. Proof of this contention is the good global correlation of chemical weathering of silicates with physical erosion (Gaillardet et al., 1999).

The idea that past Himalayan uplift resulted in increased weathering on a global scale has been promoted by the study of strontium isotopes. Late Cenozoic seawater is notable for a sharp rise in the $^{87}Sr/^{86}Sr$ ratio as recorded by dated carbonate rocks. Principal sources of Sr to the ocean include input from rivers from continental weathering and deep ocean basalt–seawater reaction. Continental rocks are on the average higher in $^{87}Sr/^{86}Sr$ than are submarine basalts. Thus, it has been hypothesized that the increase in oceanic $^{87}Sr/^{86}Sr$ during the late Cenozoic was due to globally increased rates of weathering and input of strontium from the continents due to mountain uplift. Because most Sr occurs substituted for Ca in minerals, past Sr weathering fluxes presumably can be related to past Ca fluxes and rates of uptake of CO_2 via the weathering of Ca silicates.

Quantitative estimates of the increase in silicate weathering rate due to Himalayan uplift have been made by Richter et al. (1992) based on the marine Sr isotopic record. However, changes in the $^{87}Sr/^{86}Sr$ value for the ocean can also be due to changes in the average $^{87}Sr/^{86}Sr$ of the rocks being weathered rather than due to changes in global weathering rate. This latter conclusion has been emphasized by a number of studies (e.g., Edmond, 1992; Blum et al., 1998; Galy et al., 1999). These studies found that the rocks of the high Himalayas are exceedingly radiogenic and that much of the radiogenic Sr, as well as Ca, in Himalayan rivers is derived from the weathering of carbonates, not silicates. Because carbonate weathering, as pointed out in the Introduction, does not lead to changes in CO_2 on a multimillion-year time scale, the use of $^{87}Sr/^{86}Sr$ to deduce changes in weathering rates has fallen into general disfavor, along with the idea that the Himalayas played a role in bringing about a late Cenozoic drop in CO_2 (e.g., Blum et al., 1998; Jacobson et al., 2003).

However, the Himalayas and other mountain chains still must have some importance to global weathering and the long-term carbon cycle. Once corrections for carbonate weathering are made, the chemical weathering rate of silicates can be deduced, and this has been done for two small Himalayan watersheds (West et al., 2002). West et al. found that in the high Himalayas silicate weathering is low and dominated by carbonate weathering, as found by the studies cited above, but in the Middle Hills and Ganges Basin silicate weathering is rapid and, on an areal basis, equivalent to other areas noted for their rapid silicate weathering rates. The high weathering rate is ascribed by West et al. to the input of fresh eroded material from the high Himalayas to the hot, wet, and heavily vegetated foothills. Perhaps the major role of high mountains at low latitudes, such as the Himalayas, is to provide abundant, physically eroded, fresh bedrock material to weathering at lower elevations.

Another area of the world where mountain uplift during the Miocene may have had a major effect on atmospheric CO_2 due to enhanced weathering is the exhumation of the northern New Guinea arc terrain (Reusch and Maasch, 1998). The uplift of a volcanic arc to become a mountain belt would result in a change from net CO_2 release to the atmosphere via volcanism to net uptake via weathering of the volcanics and associated sediments. Weathering in this area would have been accelerated by the warm, wet climate at that time (and at present). Basalts, the major volcanic rock type in this situation, would weather more rapidly than the more granitic composition of the Himalayas (see "Lithology and Weathering" in this chapter).

A more serious problem with strontium isotope modeling, as applied to the mountain uplift hypothesis for atmospheric CO_2 control, has been ignoring mass balance in the carbon cycle. Raymo (1991) and Richter et al. (1992), based on Sr isotope modeling, call for an increase in rates of CO_2 uptake by silicate weathering during a period when there was no known increase in rates of CO_2 supply to the atmosphere by volcanic/metamorphic degassing. As pointed out by Kump and Arthur (1997) and Berner and Caldeira (1997), because there is so little CO_2 in the atmosphere (and oceans with which it exchanges carbon), an excess of atmospheric output over input leads to a rapid drop of CO_2 to zero in less than a million years. If mountain uplift leads to increased atmospheric CO_2 uptake, with no accompanying increased input to the atmosphere from volcanism, then another counterbalancing process with decreased uptake is necessary. The simplest counterbalance is a deceleration globally of weathering due to lower temperatures accompanying lower CO_2 levels. In this way the actual *rate* of CO_2 uptake by weathering does not change; it is controlled by the rate of emission of CO_2 to the atmosphere. Instead, acceleration due to uplift is balanced by deceleration due to global cooling, and atmospheric carbon mass balance is maintained (Berner, 1991,1994; Kump and Arthur, 1997;

Francois and Godderis, 1998). In this way the atmospheric greenhouse effect (see next section) serves as a negative feedback for stabilizing CO_2 and climate against possibly large perturbations such as mountain uplift.

Nonetheless, it is still possible to use strontium isotope in a long-term carbon cycle model. One approach (Berner and Rye, 1992) is to ensure carbon mass balance while letting the variation of $^{87}Sr/^{86}Sr$ be due to changes in the relative proportions of granite (high $^{87}Sr/^{86}Sr$) weathering versus basalt (low $^{87}Sr/^{86}Sr$) weathering on the continents. In this case variations in oceanic $^{87}Sr/^{86}Sr$ do not represent changes in anything other than the source of the strontium. This approach has been suggested recently by studies that emphasize the quantitative importance of basalt weathering over time (Dessert et al., 2003).

Another approach is to assume that the $^{87}Sr/^{86}Sr$ of rocks undergoing weathering on the continents varies with time without specifying the rock types and that higher $^{87}Sr/^{86}Sr$ implies faster weathering. The idea is that old, highly radiogenic rocks are characteristic of the cores of orogenic mountain belts and that the highly radiogenic signature is a sign of increased weathering due to mountain uplift and the exposure of the old radiogenic rocks to weathering. In this way there is a loose connection between the Sr cycle and the C cycle without imbalances in either. I adopted this approach (Berner, 1994) in terms of an adjustable correlation parameter between $^{87}Sr/^{86}Sr$ and the acceleration of weathering. The appropriate expression derived in that study is:

$$f_R(t) = 1 - L\,[(R_{ocb}(t) - R_{ocm}(t))/(R_{ocb}(t) - 0.700)] \qquad (2.1)$$

where

$f_R(t)$ = dimensionless parameter expressing the effect of mountain uplift on CO_2 uptake by the weathering of Ca and Mg silicates

$R_{ocm}(t)$ = measured $^{87}Sr/^{86}Sr$ value of the oceans as recorded by limestones

$R_{ocb}(t)$ = calculated $^{87}Sr/^{86}Sr$ value for the oceans for submarine basalt–seawater reaction alone

L = adjustable empirical parameter expressing the effect of $^{87}Sr/^{86}Sr$ on weathering rate.

The values of $R_{ocb}(t)$ were calculated from changes in the rates of basalt–seawater reaction assuming that they directly follow changes in rates of seafloor production (for a detailed discussion of seafloor production and spreading rate, see chapter 4). Excess values of measured $^{87}Sr/^{86}Sr$ over the calculated $R_{ocb}(t)$ values are assumed in equation (2–1) to reflect increased input of radiogenic Sr from the continents. The parameter L is varied in the GEOCARB carbon cycle model (Berner, 1994) to investigate sensitivity of Sr isotope variation to calculated atmospheric CO_2 level.

A more direct approach to mountain uplift and weathering is to estimate actual rates of continental erosion over geologic time. This has been done recently and incorporated into long-term carbon cycle models (Berner and Kothavala, 2001; Wallmann, 2001). A direct measure of global erosion in the geological past is the global rate by which siliciclastic rocks (sandstones and shales) were deposited. Such rocks, by definition, are derived by physical erosion. The abundance of sandstones plus shales over the Phanerozoic has been estimated from studies of thousands of rock occurrences by Ronov (1993). The data, listed for ages going back every 10–30 million years, has been fitted with an exponential decay relation representing, as a first-order approximation, the loss of the rocks as a result of their own erosion (Wold and Hay, 1990). The proper expression for this is:

$$\Delta V/\Delta t = (\Delta V/\Delta t)_o \exp(-k\tau) \qquad (2.2)$$

where

τ = mean age of a given volume of rocks deposited over the time span $\Delta\tau$
ΔV = volume of rocks of age span Δt
$(\Delta V/\Delta t)_o$ = rate of deposition at "present"
k = erosion decay coefficient (assumed constant).

Values of $\Delta V/\Delta t$ are determined from the data of Ronov (1993) for 27 age spans ranging from the Miocene to the lower Cambrian. A plot fitted to the Ronov data is shown in figure 2.1. To avoid overly biasing the fitted curve by including excessive erosion accompanying the Plio-Pleistocene glaciation, "present" is assumed to represent the mid-Miocene (15 Ma). Also, use of a constant value for k assumes that the probability of erosive loss does not change with time. Deviations of each time span data point above and below the exponential curve can then be interpreted as original increases or decreases in global sedimentation (i.e., erosion) rate relative to that at present (Wold and Hay, 1990). In other words,

$$R_{depn}(t) = [\Delta V/\Delta t_{ron}/\Delta V/\Delta t_{exp}] R_{depn}(o) \qquad (2.3)$$

where $R_{depn}(t)$ = rate of deposition at time t, and the subscripts ron and exp refer, respectively, to the measured values of Ronov and the expected values for each time for simple exponential decay. The symbol (o) refers to the Miocene present. This expression can be recast in terms of a dimensionless parameter:

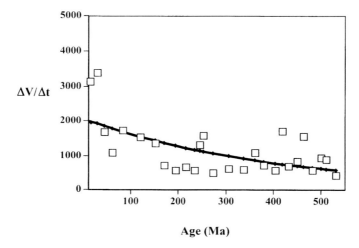

Figure 2.1. Exponential fit to the data of Ronov (1993) for the volume of sandstones and shales per unit time plotted against geological age. (Modified from Berner and Kothavala, 2001.)

$$f_{erosion}\,(t) = R_{depn}(t) \;/\; R_{depn}(o) \qquad (2.4)$$
$$= [\Delta V/\Delta t_{ron}/\Delta V/\Delta t_{exp}]$$

Finally, based on the findings of a log-log linear relation between physical erosion rate and silicate weathering rate for the world's major rivers (Gaillardet et al., 1999), one can express the relative effect of mountain uplift and erosion on the rate of chemical weathering of Ca and Mg silicates as the dimensionless parameter:

$$f_R(t) = [f_{erosion}\,(t)]^{2/3} \qquad (2.5)$$

Equation (2.5) is another derivation for the parameter $f_R(t)$ (see equation 2.1). A cubic fit to values of $f_R(t)$ over Phanerozoic time is shown in figure 2.2. Included also in figure 2.2 are the results for $f_R(t)$ calculated from strontium isotopes via equation (2.1) using L = 2. It is interesting that isotope-based $f_R(t)$ values are in very good agreement with the curve fit to the sediment data. This agreement between two independent methods suggests that the use of Sr isotopes to describe the effect of mountain uplift on silicate chemical weathering has some validity.

Plants and Weathering

There is little doubt that land plants accelerate the chemical weathering of silicate minerals (for a summary, consult Berner et al., 2003). This is accomplished in a variety of ways. First, rootlets (+ symbiotic micro-

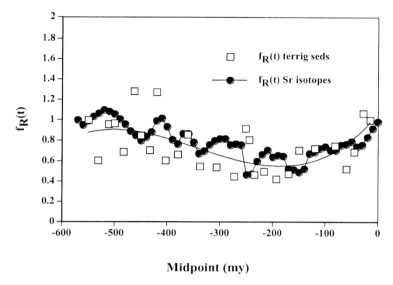

Figure 2.2. Plot of $f_R(t)$ versus time based on Sr isotope modeling and on the abundance of Phanerozoic terrigenous sediments (sandstones and shales). The curve is a cubic fit to the sediment abundance data. The parameter $f_R(t)$ is a measure of the effect of physical erosion on Ca and Mg silicate weathering. (Modified from Berner and Kothavala, 2001.)

flora such as mycorrhizae) secrete organic acids and chelates that attack primary minerals in order to gain nutrients. The principal weathering-supplied nutrients are Ca, K, and Mg. The nutrients that are liberated from minerals can be taken up by the growing plants or lost from the soil via drainage. When the plants eventually die, part of their nutrients are taken up by new plants, but part is also lost in the drainage. Over time the biogeochemical cycling of the nutrients results in loss to streams and eventually to the ocean. As a result, land plants accelerate the transfer of dissolved Ca and Mg, and associated HCO_3^-, from the continents to the oceans, leading to an increased uptake of atmospheric CO_2 during weathering (reaction 1.1).

Of special interest is the liberation of nutrients from specific minerals. Field observations (April and Keller, 1990; Griffiths et al., 1996; Berner and Cochran, 1998; van Breeman et al., 2000) show that roots and the hyphae of mycorrhizae can penetrate into rocks and selectively dissolve those minerals rich in Ca and K. This is accomplished by the secretion of organic acids, which supply both hydrogen ions and chelating agents (such as oxalate) for the complete dissolution of the minerals. Complete dissolution is shown by molds of preexisting crystals. Selective biologically induced dissolution of a normally less reactive mineral (plagioclase) in a normally more reactive matrix (volcanic glass) is illustrated in figure 2.3. This shows the importance of vegetation in

Figure 2.3. Dissolution of a plagioclase phenocryst in basalt by a presumed fungal hypha. The entry tube formed by the hypha to the phenocryst is open to the outside of the rock, which was originally connected to a plant root that was accidentally removed during sample preparation. The plagioclase is preferentially dissolved, presumably because of its higher Ca content, relative to the surrounding glassy matrix. (Modified from Berner, 1995.)

bringing about silicate weathering (similar photomicrographs can be found in Berner and Cochran, 1998).

Land plants can bring about enhanced weathering in additional ways. Organic litter accumulates in soil and undergoes microbial decomposition to organic acids and carbonic acids, which provide additional H⁺ and chelates for mineral dissolution. Plants recirculate water via evapotranspiration. For example, much of the rainfall in heavily forested regions, such as the Amazonian lowlands, is formed from the condensation of recirculated water transpired from the trees (Shukla and Mintz, 1982). In this way the trees act as natural soxhlet extractors whereby dilute recirculated water is constantly added to the soil for mineral dissolution.

It is often stated that plant roots protect land from erosion, and this can be interpreted as inhibiting chemical weathering by keeping underlying bedrock minerals from being exposed to weathering. However, this applies only to areas where all primary minerals have been removed from the soil by weathering and removal of the soil cover is limited. In areas of moderate slope, plants hold soil against erosion and allow moisture adsorbed on clays to build up, which enables continued dissolution of primary minerals still disseminated in the soil (Drever, 1994). In the absence of vegetation, rapid erosion can expose bare bedrock, which holds less moisture, so that less weathering takes place because of shorter water-rock contact time.

Although much has been learned about plants and weathering by studying modern ecosystems (e.g., Likens et al., 1977), what is needed in studying the Phanerozoic carbon cycle is the quantitative significance of plants as they affect CO_2 uptake during weathering. Vascular plants invaded the continents during the early Paleozoic (Gensel and Edwards, 2001), but it was not until the Devonian that large plants with deep roots, such as trees, became really important. Large vascular plants must have weathered rocks faster than the algae, lichens, or bryophytes that preceded them (Berner, 1998). This is because trees have vast rootlet systems that expose a large interface between roots and minerals, allowing for the rapid uptake of nutrients to form large, fast-growing bodies. In contrast, even though there is evidence that they weather minerals (Barker and Banfield, 1996; Aghamiri and Schwartzman, 2002), lichens are small, have a small interface with rocks, and grow very slowly. Thus, for algae, lichens, and bryophytes there must be slower biogeochemical nutrient cycling and, thus, slower weathering. The rarity of soils developed under lichens, as compared to those under trees, attests to the efficacy of trees in accelerating weathering to a much greater extent.

The quantitative effect of trees, relative to mosses and lichens, on rates of weathering has been estimated by field experiments (table 2.1). This is a difficult task because it is necessary to hold constant all other factors that affect weathering, such as climate, relief, and lithology, to discern the vegetational effect. As a result there are only a few studies of this sort. Drever and Zobrist (1992) used water chemical analyses to examine the rate of release of HCO_3^- from watersheds of relatively uniform granitic lithology and relief at different elevations in the southern Swiss Alps. They found that the weathering flux was about 24 times higher under forested land than that above the tree-line (where lichens and bryophytes are present). After correcting for the change of temperature with elevation as it affects weathering rate, their results show that plants accelerate CO_2 uptake during weathering by a factor of about 8.

Table 2.1. Ratio of weathering fluxes from vegetated areas versus areas sporadically covered by mosses and lichens.

	HCO_3	Mg	Ca	Si	Reference
Iceland[a]					
Birch/moss	3	4	3	2	Moulton et al., 2001
Conifer/moss	3	4	3	3	Moulton et al., 2001
Southern Swiss Alps[b,c]	8	—	—	8	Drever and Zobrist, 1992
Colorado Rocky Mts.[b]	Na + K + Mg + Ca = 4				Arthur and Fahey, 1993

[a]Tree storage plus runoff.
[b]Runoff only.
[c]Corrected for temperature by elevation.
A dash means no data.

The effect on release rate of total cations in drainage, by the presence and absence of forests on adjacent land areas, has been studied by Arthur and Fahey (1993) at a high-elevation granitic site in Colorado, USA, and their results indicate acceleration by the trees of a factor of about 4.

The effect of trees versus mosses and lichens during basalt weathering has been studied by Moulton et al. (2000) in Iceland. Iceland was chosen for this work because small areas of forested and nonforested land were found adjacent to one another on the same rock type under the same relief, elevation, and microclimate. Also, Iceland receives no anthropogenic acid rain, and basalt weathering is especially important to the long-term carbon cycle because it consists of highly weatherable Ca and Mg silicate minerals (Dessert et al., 2001). Moulton et al. looked at the drainage release and storage in trees of Ca and Mg and found an acceleration of weathering by a factor of 3–4 over the adjacent moss- and lichen-covered ground (table 2.1).

The data of Millot et al. (2003) suggest that there is an enhancement of silicate weathering rate by vegetation in the Mackenzie River basin of Canada. They found that the composition of stream waters from the lowland plains are enriched in dissolved organic matter compared to the mountainous headwaters and that the silicate weathering rates of the plains rivers are 3–4 times faster than the mountain rivers. (Because both areas are forested, their results are not listed in table 2.1.) Millot et al. explain this difference by weathering rates to the higher content of chelating organic compounds in the less well-drained organic-rich soils of the plains. The results of Millot et al. are of special interest to the study of the effect of plants on weathering because they show that differences in vegetation can complicate any simple correlation between relief and silicate weathering rate.

A criticism is sometimes offered that weathering studies of young growing forests, as listed in table 2.1, do not take into consideration the ultimate attainment of thoroughly leached soils under old forests. Chadwick et al. (1999) have shown that after about 20,000 years of weathering, basaltic rocks in the Hawaiian Islands are leached of all nutrients, with the result that trees no longer weather bedrock and become dependent on rainfall and atmospheric dust for a continued supply of nutrients. Likewise, in the Amazonian lowlands, soils are so thick and thoroughly leached that bedrock weathering has essentially ceased, and the only supply of nutrients is that derived from the recycling of forest litter and dead trees (Stallard and Edmond, 1983). These are valid criticisms, but they only apply to flat ground. (The Hawaiian study examined only soils on areas with low slopes.) Most weathering takes place on hillslopes where there is sufficient gradient for removal of clay weathering products by physical erosion. In areas of high rainfall and high slope, landsliding is frequent and common (Stallard, 1995; Hovius et al., 1997), and this results in the uprooting and destruction of vegetation as well as the uncovering of fresh bedrock for continued weathering. Following landslides new trees are reestablished, and if the landsliding is frequent enough, the state of old forests with completely leached soils is never attained.

It is often misconstrued that an increase in the rate of silicate weathering brought about by the rise of large vascular plants must have led to the production globally of greater quantities of soil clays and that this hypothesis could be checked by examining how the composition and abundance of clay minerals may have varied over the Phanerozoic (e.g., Algeo and Scheckler, 1998). This reasoning is incorrect because of the necessity of balancing the carbon cycle. The increased uptake of atmospheric CO_2 by vegetation-assisted weathering must have been balanced either by greater inputs of CO_2 to the atmosphere from volcanic/metamorphic degassing and/or by other counterbalancing processes that decelerate the rate of weathering. As pointed out earlier, there is very little CO_2 in the atmosphere–ocean system, and a small imbalance in the rates of CO_2 removal and addition can lead to loss of all atmospheric CO_2 in less than 1 million years (Berner and Caldeira, 1997). There is no evidence of an increase in degassing during the rise of the large vascular plants, so there must have been a deceleration of weathering due to another process or processes. The likely deceleration process was a drop in CO_2 leading to a cooler earth via the atmospheric greenhouse effect. In this way the acceleration of weathering by plants was almost totally balanced by the deceleration of weathering by a drop in global mean temperature. The drop in CO_2 was brought about by plant-assisted weathering, but it occurred only as a small leak in an otherwise well-balanced carbon cycle. After the rise of the large plants, a new steady state between atmospheric CO_2 inputs and outputs was achieved at a lower CO_2 level, but during the drop in CO_2 the actual rate of weathering could have remained almost constant.

Of further interest to the Phanerozoic carbon cycle is the rise of angiosperms during the Cretaceous. How did this affect rates of plant-assisted weathering? Studies of modern ecosystems do not provide a clear answer. Moulton et al. (2000) found that in Iceland, the weathering of basalt was 50% faster for angiosperms (dwarf birch) than for gymnosperms (conifers) when normalized to biomass. Red alder (angiosperm) forest in the western Washington Cascade Mountain foothills loses Ca and Mg eight times faster than adjacent Douglas-fir forest (gymnosperms) of similar age and biomass (Homann et al., 1992). In contrast, Quideau et al. (1996), studying two experimental ecosystems in southern California, concluded that gymnosperms (pine) release Ca and Mg from primary minerals faster than angiosperms (scrub oak). In an area of northern Minnesota, Bouabid et al. (1995) found that plagioclase, of fixed Ca/Na composition, exhibited approximately equal degrees of surface etch pitting in soils underneath stands of pine and oak-basswood. Augusto et al. (2000) inserted weighed mineral samples under a variety of stands of confers and hardwoods of northern France and found that after 9 years there was a distinctly greater mass loss of plagioclase

under the conifers relative to the hardwoods. These results point to a
need for further study using carefully defined experiments before anything can be said about how the rise of angiosperms may have affected
the long-term carbon cycle.

In GEOCARB long-term carbon cycle modeling (Berner, 1991, 1994;
Berner and Kothavala, 2001), the effect of plant evolution is expressed in
terms of the dimensionless parameter $f_E(t)$. The value of $f_E(t)$ is set equal
to 1 for the present, representing angiosperm-dominated weathering. For
the prevascular land-plant surface, a value <1 is chosen based on the field
studies discussed above (for example, an acceleration of weathering of a
factor of 8 would be expressed by $f_E(t) = 0.125$). For the transition from a
lichen/bryophyte world to that dominated by large vascular plants, the
value of $f_E(t)$ is assumed to rise linearly with time for about 30 million
years (380–350 Ma), reaching a new value at 350 Ma, characteristic of
weathering by gymnosperms (plus more primitive vascular plants). During the Cretaceous, a 50 million-year transition (130–80 Ma) from the
gymnosperm $f_E(t)$ value to $f_E(t) = 1$ for a angiosperm-dominated world is
assumed. After 80 Ma, $f_E(t)$ is assumed to stay constant at 1 (these ages
are in agreement with the latest paleobotanical literature; see Willis and
McElwain, 2002).

Plant-assisted weathering can respond to changes in atmospheric CO_2
because plants grow faster at higher CO_2 levels. Many laboratory experiments (e.g., Bazzaz, 1990) have shown that plants fix more carbon at
elevated CO_2 if growth is not limited by water, nutrients, or light. If plants
grow faster, they must take up nutrients faster, and, thus, weather rocks
faster. In this way there exists a biological negative feedback effect of
changes in CO_2 on weathering rate. Evaluation of this effect for a natural forest has been done by Andrews and Schlesinger (2001). They irrigated portions of a North Carolina pine forest with elevated levels of
CO_2 and compared the flux of dissolved bicarbonate in soil waters under the irrigated and nonirrigated pine trees. They found a 33% increase
in HCO_3^- release (in other words, weathering rate) for a change of atmospheric CO_2 from 360 ppm to 570 ppm.

This biological negative feedback effect has been parameterized in
GEOCARB modeling (Berner, 1994; Berner and Kothavala, 2001) by the
dimensionless parameter $f_{Bb}(CO_2)$ expressed as the Michaelis-Menton
equation:

$$f_{Bb}(CO_2) = [2RCO_2/(1 + RCO_2)]^n \qquad (2.6)$$

where

RCO_2 = mass of carbon dioxide in the atmosphere at some past time
divided by the mass at the preindustrial present (280 ppm)

n = exponent representing the efficacy of CO_2 in fertilizing
plant growth globally.

The exponent n is used to indicate that plant growth at many localities is not affected by changes in CO_2 because of limitation of growth by nutrients, water, or light. The value n = 0 means no fertilization globally, whereas a value of n =1 means that all plants globally respond to CO_2 fertilization. A big problem, affecting both the modelling of the long-term carbon cycle and future predictions of rises in atmospheric CO_2, is the value of n. A Michaelis-Menton formulation is used to express that there is a limit to plant productivity, and thus a response to increasing CO_2, on land. Other more complex formulations have been offered (e.g., Volk, 1989), but equation (2.6) is used as a simple first approximation for a process that is poorly understood.

Before the rise of large vascular land plants, there still must have been a negative feedback for stabilizing atmospheric CO_2. One such feedback is the atmospheric greenhouse effect discussed in the next section. The other is the direct effect of atmospheric CO_2 on weathering. In the absence of plants, increases in atmospheric CO_2 would still result in faster weathering as CO_2-enriched rain fell onto the land and additional CO_2 diffused from the atmosphere into the soil. The principal weathering agent in both cases would be carbonic acid. The buildup of high levels of CO_2 and carbonic acid in soils at present, with diffusion out to the atmosphere, is due to root respiration and the microbial decomposition of organic matter. This would not happen on a biota-free land surface. Such a simple situation of CO_2-rich rain falling on the land or CO_2 diffusing from the atmosphere into the soil and reacting with silicate minerals is represented by laboratory mineral dissolution experiments (e.g., see Lasaga, 1998, for a summary). If the kinetics of these experiments are assumed for the biota-free situation, then an appropriate dimensionless weathering rate expression is (see Berner, 1994, for further details):

$$f_{nBb} (CO_2) = [RCO_2]^{0.5} \tag{2.7}$$

where $f_{nBb}(CO_2)$ is the weathering rate for a biota-free land surface divided by the same rate for a biota-free surface with the present level of atmospheric CO_2. The subscript nBb refers to no biology. The question remains as to the applicability of equation (2.7) to a land surface populated by algae, lichens, and/or bryophytes. More experimental work is needed to discern the response of weathering brought about by these primitive organisms to changes in CO_2. In the absence of such data, GEOCARB modeling assumes, as a first-order approximation, equation (2.7).

Atmospheric Greenhouse Effect and Weathering

Changes in the concentration of greenhouse gases affect both the temperature and the hydrology of the continents, which in turn affect the rate of uptake of CO_2 via silicate mineral weathering. The principal greenhouse gases of interest are CO_2 and CH_4. (Although H_2O is the strongest greenhouse gas, it is buffered by evaporation and condensation that is driven by external factors such as solar radiation and the CO_2 greenhouse effect.) The buildup of CO_2 in the atmosphere can lead to higher temperatures, more rain on the continents, more runoff, and thus faster weathering. It is well established that minerals dissolve faster at higher temperatures and with greater rainfall (e.g., Jenny, 1941). Thus, changes in weathering rate induced by variations in CO_2 can serve as a negative feedback for stabilizing global temperature (Walker et al., 1981; Berner et al., 1983). This is illustrated in figure 2.4 in terms of a simple systems analysis feedback diagram.

The effect of changes in concentrations of methane can be important to weathering only when it becomes the dominant greenhouse gas. This was probably the case for the Archean (Pavlov et al., 2000) and possibly much of the Proterozoic (Schrag et al., 2002; Pavlov et al., 2003). However, for the Phanerozoic this was unlikely because of the presence of relatively high levels of O_2 (compared to the Precambrian). Because CH_4 is rapidly oxidized to CO_2 in the atmosphere (residence time of atmospheric CH_4 is only about 10 years), Phanerozoic levels of CH_4 probably never were high enough over sufficiently long periods to act as the dominant greenhouse gas.

Results of general circulation models (GCMs) for global mean surface temperature versus CO_2 concentration can be represented rather well by the simple expression (Berner, 1991):

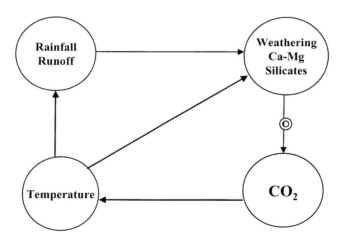

Figure 2.4. Systems analysis diagram for the greenhouse-silicate weathering feedback. In such diagrams arrows with bullseyes represent negative response; without arrows positive response. A complete cycle with an odd numbers of bullseyes means negative feedback and stabilization; a complete cycle with an even number of bullseyes, or no bullseyes, means positive feedback and (normally) destabilization.

$$T(t) - T(0) = \Gamma \ln RCO_2 \qquad (2.8)$$

where

T (t) = global mean surface temperature at some past time
T(0) = global mean surface temperature for the present
RCO_2 = ratio of mass of CO_2 in the atmosphere at time t to that at
present
Γ = coefficient derived from GCM modeling.

For the purpose of studying the carbon cycle on a Phanerozoic time scale, the "present" can be assumed to be preindustrial with a CO_2 concentration of 280 ppm and a global mean temperature of 15°C.

The effect of temperature on the rate of primary mineral dissolution during weathering can be deduced both from laboratory and field studies. To represent the temperature effect, it is common to employ the "activation energy," which practically is a measured temperature coefficient. The so-called Arrhenius expression used for this purpose is:

$$\ln (J / J_o) = (\Delta E/R)(1/T_o - 1/T) \qquad (2.9)$$

where

T = absolute temperature in degrees K
T_o = absolute temperature for some standard state (here assumed
to be 288°C the present global mean surface temperature
ΔE = activation energy
R = gas constant
J = dissolution rate in terms of an equivalent weathering uptake
of CO_2 to form dissolved HCO_3^-. Units are mass per unit volume of soil (regolith) water per unit time
J_o = dissolution rate for the standard state.

The results of laboratory studies on silicate dissolution (Brady, 1991; Blum and Stillings, 1995; White et al., 1999), in terms of activation energy, are shown in table 2.2. For mathematical convenience equation (2.9) can be rewritten as:

$$\ln (J/J_o) = (\Delta E/R) (T - T_o)/TT_o \qquad (2.10)$$

Because T and T_o are large numbers that are rather close to one another at earth surface temperatures, their product can be considered as essentially constant, so that, solving for J,

$$J /J_o = \exp [Z(T - T_o)] \qquad (2.11)$$

Table 2.2. Activation energies for silicate rock and mineral dissolution.

Rock or mineral	ΔE (kJ/mol)	Reference
Laboratory experiments		
Olivine	38–86	Brady, 1991 (compilation)
Enstatite	41–80	Brantley and Chen, 1995
Diopside	50–150	Brady, 1991 (compilation)
Wollastonite	72	Brady, 1991 (compilation)
Na-plagioclase	60	Blum and Stillings, 1995
K-feldspar	52	Blum and Stillings, 1995
Granite/granodiorite	47–60[a]	White et al., 1999
Granite/granodiorite	53–71[b]	White et al., 1999
Field studies		
Plagioclase	77	Velbel, 1993
Plagioclase	97	Brady et al., 1999
Plagioclase	55[c]	Brady et al., 1999
Olivine	89	Brady et al., 1999
Olivine	48	Brady et al., 1999
Granite/granodiorite	51[a]	White et al., 1999
Basalt	42	Dessert et al., 2001

[a]Based on silica dissolution.
[b]Based on Na dissolution.
[c]Mediated by lichens.

where Z is equal to $\Delta E/RTT_0$ (symbolized as ACT in Berner and Kothavala, 2001). Equation (2.11) is a form that is especially useful in carbon cycle modeling.

Field studies (Velbel, 1993; Brady et al., 1999; White et al., 1999; Dessert et al., 2001) have shown that for a given rock type one can discern a temperature effect on weathering rate providing that variations in relief and other factors are limited. Results of these studies have been summarized in terms of activation energies and are also shown in table 2.2. In general there is agreement between field studies and experimental studies indicating that the rate-limiting step in the dissolution of the primary minerals is the same in the field as in the lab. Judging by the rather high values for ΔE, this must involve reactions at the mineral surface and not diffusion of dissolved species to and from the surfaces. (For a detailed discussion of mineral dissolution mechanisms, see Lasaga, 1998.)

To convert weathering fluxes J to riverine fluxes of HCO_3^- to the oceans, some additional calculations are necessary. First, we need to know the global mean concentration of dissolved HCO_3^- in river water derived from Ca and Mg silicate weathering. Following Berner (1994), we assume, as a first approximation, that for a "global regolith" at steady state,

$$J = (rA/V) C \qquad (2.12)$$

where

C = global average concentration of dissolved HCO_3^- from silicate weathering, in mass per unit volume, in the regolith pore solution and eventually in river water

r = global mean runoff (volume per unit time per unit area)

A = surface area of that portion of the continents undergoing silicate weathering

V = volume of water contained in the global regolith.

Now

$$V = hA\Phi \qquad (2.13)$$

where h is the mean thickness of the global regolith, and Φ is its mean porosity. Combining (2.12) and (2.13) and solving for C,

$$C = h\Phi J/r \qquad (2.14)$$

Now, it has been shown (Berner, 1994) from a variety of field studies of rivers that

$$C = kr^{-0.35} \qquad (2.15)$$

This demonstrates the effect of dilution as a result of increasing runoff. The same form of this expression can be independently derived from equation (2.14) (Berner, 1994), resulting in

$$C = \Phi k'J r^{-0.35} \qquad (2.16)$$

where k' is a parameter expressing the relation between the mean thickness of the global regolith and runoff. Equation (2.16) is used to convert J to C to determine the global flux of HCO_3^- from silicate weathering.

In doing carbon cycle modeling, it is important to remember that it is the temperature of the land actually undergoing weathering that is relevant. Thus, the use of global mean temperature (which includes the oceans) as it relates to CO_2 level (equation 2.8) is an oversimplification. However, in the absence of available paleo-land temperature versus CO_2 data, modeling to date has been forced to use this simplification (e.g., Walker et al., 1981; Berner, 1994; Wallmann, 2001). Furthermore, the mean temperature of the land is inappropriate because it includes areas of glaciers or deserts where there is virtually no chemical weathering. An attempt to apply a more rigorous approach, by looking at the relation between CO_2 and the temperature of land with precipitation >25 cm/year (no deserts) and a mean yearly temperature >−5°C (no glaciers), has been applied to a GCM study of weathering during the Cretaceous (80 Ma) by Kothavala, Grocke, and Berner (unpublished ms).

Besides temperature, rainfall and the flushing of the regolith is also important in weathering. With all other factors held constant, flushing can be represented by runoff from the land. Runoff is affected by changes in both local and global climate. Local climate is a function, on a geological time scale, of continental drift as land areas pass from dry to wet climatic zones. This effect of paleogeography on runoff will be discussed in the next section. Concern here is with the effect of changes in global mean temperatute, due to the greenhouse effect, on runoff. The relation between global mean temperature and runoff can be deduced from GCM models, but to my knowledge this relation has not been calculated for the distant past with paleogeographies different from that at present. Using present geography, Berner and Kothavala (2001) deduced, on the basis of GCM modeling, the expression:

$$r(T)/r(T_o) = 1 + Y(T - T_o) \qquad (2.17)$$

where r represents runoff, T is global mean temperature at some past time, and T_o is that for the preindustrial present. Y is an empirical parameter fit to the GCM results (symbolized as RUN in Berner and Kothavala, 2001).

Calculation of the global weathering uptake of atmospheric CO_2 and riverine flux of HCO_3^- to the oceans is done according to

$$\text{Flux} = C(T)\, r(T) \qquad (2.18)$$

where flux is in mass per unit of land area. To normalize weathering to that at present, we have

$$f_B(T) = [C(T)/C(T_o)]\, [r(T)/r(T_o)] \qquad (2.19)$$

where $f_B(T)$ equals flux (T)/flux (T_o), the dimensionless parameter expressing the effect of global mean temperature on the uptake of CO_2 to form dissolved HCO_3^- via the weathering of silicates. Assuming that the parameters Φ and k' do not change with temperature, we obtain from equations (2.16) and (2.19):

$$f_B(T) = [J(T)/J(T_o)]\, [r(T)/r(T_o)]^{0.65} \qquad (2.20)$$

To combine the effects of temperature on runoff with that on dissolution rate, we obtain from equations (2.17) and (2.20):

$$f_B(T) = [J(T)/J(T_o)] \ [1 + Y(T - T_o)]^{0.65} \tag{2.21}$$

which on substituting equation (2.11) yields

$$f_B(T) = \exp[Z(T - T_o)] \ [1 + Y(T - T_o)]^{0.65} \tag{2.22}$$

Finally, to obtain the greenhouse effect of CO_2 on the rate of silicate weathering uptake of CO_2 to form HCO_3^-, we combine equation (2.22) with the GCM greenhouse equation (2.8) to obtain the nondimensional greenhouse parameter $f_{Bg}(CO2)$ (subscript g stands for greenhouse):

$$f_{Bg}(CO_2) = (RCO_2{}^{Z\Gamma}) \ (1 + Y\Gamma \ln RCO_2)^{0.65} \tag{2.23}$$

A plot of equation (2.23), which represents the greenhouse-caused negative feedback response to changes in atmospheric CO_2, is shown in figure 2.5.

Solar Radiation, Cosmic Rays, and Weathering

As emphasized in the previous section, the rate of mineral dissolution during weathering is a function of the temperature of the land. Besides the atmospheric greenhouse effect, there is also the effect of changes in solar radiation on both global and land surface temperature. It is well documented by solar physicists (Endal and Sofia, 1981; Gough, 1981) that the gradual evolution of the sun over geologic time has resulted in increasing levels of radiation reaching the earth. On a geologic time scale the effect has been dramatic. At 4 Ga the level of radiation is estimated to have been 30% less than today, which means that the oceans should have been completely frozen. However, the presence of water-lain sediments at that time indicates that some other warming process must have been present to avoid global oceanic freezing. The generally agreed upon culprit is a very strong atmospheric greenhouse effect due either to very high levels of CO_2 (e.g., Kasting and Ackerman, 1986) or of CH_4 (e.g., Pavlov et al., 2000).

Over the Phanerozoic solar evolution has continued to increase linearly, starting at a level of about 6% less than now at the start of the Cambrian. This is still a dramatic effect. The level of elevated atmospheric CO_2 necessary to counter this reduced radiation, in order to attain a global mean surface temperature the same as at present, can be calculated simply via a modification of equation (2.8):

$$T(t) - T(0) = \Gamma \ln RCO_2 - Ws(t/570) \tag{2.24}$$

where Ws expresses the effect on temperature of the linear increase in solar radiation with time. Using the values for Ws (7.4) and Γ (3.3°C)

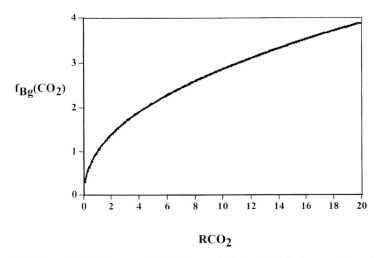

Figure 2.5. Plot of $f_{Bg}(CO_2)$ versus RCO_2. The variable $f_{Bg}(CO_2)$ is the nondimensional factor expressing the effect of temperature, as controlled by the atmospheric greenhouse effect, on silicate weathering. RCO_2 is the ratio of CO_2 concentration at some past time to that at present. The relation between the two parameters is given by equation (2.23) with $Z = 0.09$, $Ws = 7.4$, and $\Gamma = 4°C$. The reduction in slope of the curve with increasing RCO_2 reflects the fact that with increasing CO_2 levels the atmospheric greenhouse effect, per unit change in CO_2, diminishes.

for the Cambrian, based on the NCAR CCM-3 GCM model (Berner and Kothavala, 2001) indicates that to maintain the same global mean surface temperature at the beginning of the Cambrian (550 Ma) as at present would require a value of RCO_2 of 8.7, or about nine times higher than today. This illustrates the importance of including variations in solar radiation in carbon cycle modeling over long geologic times. A plot of RCO_2 versus time necessary for the temperature to be the same as at present is shown in figure 2.6.

Recently it has been suggested that the passage of the solar system through arms of the Milky Way galaxy results in increased fluxes of cosmic rays toward the earth because of a greater density of supernovas in the spiral arms (Shaviv, 2002). Reactions of atmospheric gases with cosmic rays produce ions that can serve as cloud condensation nuclei. Thus, an increase in cosmic ray bombardment of the earth could lead to greater cloudiness and global cooling. Shaviv states that the major glaciations of the Phanerozoic can be explained as occurring during periods of spiral arm passages (although his predicted large Jurassic glaciation does not exist). If this hypothesis has merit and (it needs more verification), then an additional mechanism can be called upon to affect the rate of silicate weathering. However, greater cloudiness, while leading to cooling and slower mineral weathering, could also lead to

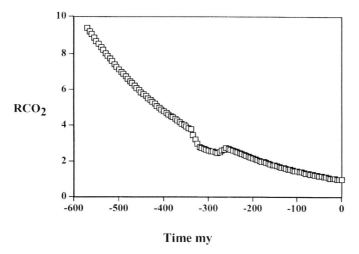

Figure 2.6. RCO_2 versus time for the situation where the increase of solar radiation with time is balanced by a diminishing CO_2 greenhouse effect to maintain mean annual surface temperature at all times to be the same as at present. Deviations from a smooth curve are due to changes in the value of the greenhouse coefficient Γ over time.

greater rainfall and faster weathering. Thus, it is not clear what the effect of changes in cosmic ray flux could have had on weathering rate.

Continental Drift: Effect on Climate and Weathering

The actual site where weathering occurs depends on the local climate. I have already shown how local climate may vary with time due to changes in global climate as a result of changes in greenhouse gas concentration, solar radiation, and so on. However, local climate can also vary because of changes in the position and size of the continents as a result of continental drift. This is a unique feature of the long-term carbon cycle. A simple argument, whereby a land mass is moved from the pole to the equator, illustrates how change in location can lead to great changes in temperature, rainfall, and the rate of CO_2 uptake by silicate weathering (Worsley and Kidder, 1991).

One approach to the problem of climate change due to changing paleogeography is to assume the same latitudinal climate zones in the past as exist at present. Then, by moving the continents across paleolatitudes, one can sum up the total evaporation and rainfall for each zonal land segment and add those for all segments to obtain total global river discharge. This has been done by Tardy et al. (1989). (A more up-to-date review of the effects of paleogography on chemical weathering is given by Tardy, 1997.)

Although the Tardy (1989) approach is a useful first-order attack on the problem of paleogeography, climate, and weathering, it misses effects due to changes in continental size and topography. Larger continents, especially those with coastal mountains, should experience more extensive monsoonal climate and have very dry interiors (rain shadows) with wet margins (orographic rainfall). This is exemplified especially by the supercontinent Pangea (e.g., Kutzbach and Gallimore, 1989). To approach all aspects of the effect of changing paleogeography on climate and weathering, it is necessary to combine paleogeographic reconstructions (e.g., Scotese and Golonka, 1995) with paleoclimate modeling (e.g., Otto-Bliesner, 1993; Barron et al., 1995; Hyde et al., 1999; Gibbs et al., 2002). For the entire Phanerozoic it would be useful to have quantitative paleoclimatic reconstructions covering most of this period, but so far this has not been done. Attention has been focused on glaciations, the Cretaceous, and the Permian and Triassic, when Pangaea was at its largest size.

Parrish et al. (1982) has attempted to estimate rainfall patterns at various times during the Phanerozoic, based on geological indicators such as coals and evaporites. However, these results cannot be used to estimate mean global runoff over time. Only a few attempts have neen made to calculate runoff (as the difference between precipitation and evporation) over appreciable portions of Phanerozoic time (Otto-Bliesner 1993, 1995; Fawcett and Barron, 1998; Gibbs et al., 1999). Only Otto-Bliesner has considered the entire Phanerozoic. She used GCM modeling and the paleogeographic reconstructions of Scotese and Golonka (1995) to calculate global mean land temperatures and mean runoff for 13 times spanning the Phanerozoic. Unfortunately, because of inadequate knowledge of paleotopography over such a long time, Otto-Bliesner was forced to assume flat, ice-free continents at sea level. Her results, as they affect the rate of chemical weathering of silicates, have been incorporated into GEOCARB modeling (Berner, 1994; Berner and Kothavala, 2001) in terms of the dimensionless parameter:

$$f_D(t) = \text{runoff}(t)/\text{runoff}(0) \qquad (2.25)$$

A plot of $f_D(t)$ versus time is shown in figure 2.7. To actually apply $f_D(t)$ to the global rate of weathering, two modifications of equation (2.25) are needed. First, the total riverine discharge of water from the continents is what is desired, and this is obtained from runnoff (which is expressed per unit land area) by $f_{AD}(t) = f_D(t)f_A(t)$ where

$$f_A(t) = \text{land area}(t)/\text{land area}(0) \qquad (2.26)$$

Second, for silicate weathering, to express the effect of dilution of dissolved HCO_3^- at elevated runoff (see equation 2.20), one should use $f_D(t)^{0.65}$. Thus, the proper term to be applied to the global weathering flux would be

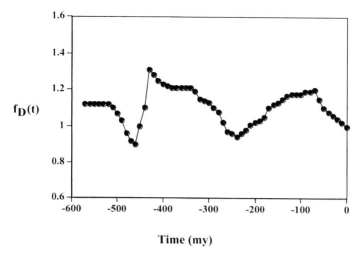

Figure 2.7. Plot of $f_D(t)$ versus time. The parameter $f_D(t)$ represents the effect of changes in paleogeography on global river runoff. Runoff is a major factor affecting silicate weathering. The variable $f_D(t)$ is defined as global river runoff at a given past time divided by that at present. High values of $f_D(t)$ represent times when a large proportion of land area was in humid belts and low $f_D(t)$ values when a large proportion was in dry belts. Data based on the GCM model of Otto-Bliesner (1995).

$$f_{AD}(t)^{0.65} = [(f_D(t)f_A(t)]^{0.65} \qquad (2.27)$$

In considering all factors affecting temperature as they relate to the rate of weathering, one should add the change in land temperature, due to changes in paleogeography, to the effects of the atmospheric greenhouse effect and solar evolution. Thus, equation (2.24) is modified to

$$T(t) - T(0) = \Gamma \ln RCO_2 - Ws(t/570) + GEOG(t) \qquad (2.28)$$

where GEOG(t) = mean land temperature at a past time minus mean land temperature at present, due solely to changes in paleogeography (in other words, for constant solar radiation and constant atmospheric CO_2 concentration). The land temperature data of Otto-Bliesner (1995) can be used for this purpose (Berner and Kothavala, 2001). As pointed out earlier, the value of GEOG(t) should be based only on land where appreciable chemical weathering can take place, which excludes deserts and areas covered by glaciers. So far this has not been done.

Equation (2.28) can be substituted in equation (2.22) to consider all factors that affect global mean temperature as it relates to weathering. This results in a more complete nondimensional expression for the effect of atmospheric CO_2 level on the rate of silicate weathering and is represented by $f_{Bt}CO_2$ (subscript t stands for temperature). By substituting equation (2.28) in (2.22), we obtain:

$$f_{Bt}(CO_2) = (RCO_2)^{Z\Gamma} \exp[ZGEOG(t) - ZWs(t/570)] \qquad (2.29)$$
$$\times \{1 + Y[\Gamma \ln RCO_2 - Ws(t/570) + GEOG(t)]\}^{0.65}$$

This is the complete expression used in GEOCARB modeling to express the effect of changes in temperature on silicate weathering.

Another aspect of paleogeography, as it has affected weathering in the past, is change in the area of land exposed to weathering, independent of climate. The original approach to carbon cycle modeling was simply that weathering rate is directly proportional to land area (Berner et al., 1983; Fischer, 1983) through the use of equation (2.26). However, this does not distinguish mountainous areas where fresh, easily weathered Ca and Mg silicate minerals are constantly exposed to weathering, from lowland areas where there is an extensive cover of relatively unreactive secondary minerals. Changes in land area over the Phanerozoic have been due largely to the transgression and regression of continental seas, and during sea-level low stands, the extra land area is underlain largely by clay weathering products (shales) and detrital primary minerals (sandstones), both of which are relatively resistant to further chemical weathering. An exception to this generalization is areas at the foot of high mountains, where rapidly eroded, unweathered debris is delivered to nearby lowlands. An outstanding modern example is the Himalayan foothills and the Indo-Gangetic plain, which receive large amounts of freshly eroded material from the Himalayas. Here, due to the presence of the unweathered primary minerals, along with warmth and high rainfall, extensive weathering takes place (e.g., West et al., 2002).

These observations make it difficult to decide whether to include change in total land area (equation 2.24) directly in carbon cycle modeling. Also, there is the problem that changes in large portions of land, present as deserts and areas covered by glaciers, should not be included in $f_A(t)$. The approach I have used so far is to exclude total land area from the expression for the rate of CO_2 uptake by the weathering of silicates (but not carbonates; see chapter 3) and to focus instead on the importance of continental relief (Berner, 1994; Berner and Kothavala, 2001).

Lithology and Weathering

The dominant source of dissolved Ca and Mg during silicate weathering is Ca contained in plagioclase (Mackenzie and Garrels, 1966; Garrels, 1967), Mg in ferromagnesian minerals such as pyroxenes, amphiboles, and biotite, and both Ca and Mg in volcanic glass. Field studies of the rate of weathering of small areas of relatively uniform rock type, subject to similar climates and relief (Meybeck, 1987; Taylor and Lasaga, 1999), have shown that basalts weather faster then granites or other acidic igneous and metamorphic rocks. This is mainly because of the presence of highly calcic plagioclase, pyroxenes, and volcanic glass in basalts, which dissolve more rapidly than the characteristic weatherable minerals of granites and gneisses, sodic/calcic plagioclase, potassic feldspars, and micas (Goldich, 1938). As pointed out in the Introduction, the weathering of Ca and Mg silicate minerals is of primary interest in the long-term carbon cycle because the released Ca and Mg ions eventually are precipitated as carbonates in the oceans, whereas Na and K do not form carbonates. Thus, basalt weathering is an important factor in controlling atmospheric CO_2. Gaillardet et al. (1999) and Dessert et al. (2003) estimate that at present 25–35% of total CO_2 uptake by silicate weathering is accounted for by basalt weathering. In addition, episodes of intense basaltic volcanism in the past could have resulted in net atmospheric CO_2 removal, rather than CO_2 addition, as normally thought (Dessert et al., 2003). This is because the consumption of CO_2 by the subsequent weathering of the basaltic minerals can be greater than the amount of CO_2 degassing accompanying the original extrusion of the basalt.

Wallmann (2001) has emphasized the importance of subaerial basalt weathering at subducting plate boundaries. There is a great degree of volcanism there, and one can assume that the degree of eruption varies in tandem with rates of subduction over geologic time. From estimates of present-day subduction-zone volcanism, the assumption that basalt weathering constitutes 25% of total CO_2 uptake by silicate weathering (Gaillardet et al., 1999), and estimates of changes over time in rates of seafloor spreading, Wallman calculated that there is an overall enhanced uptake of CO_2 by adding this process to a long-term carbon cycle model. A similar result, but of lesser magnitude, was obtained by performing the same calculation with the GEOCARB model (Berner and Kothavala, 2001).

Because calcite dissolves so much faster than any silicate mineral (White et al., 1999) and because traces of it are common in crystalline silicate rocks, such as veins and interlayers in metamorphic rocks (Ague, 2002) and inclusions in granite (White et al., 1999), chemical analyses of waters draining crystalline rocks can lead to erroneous conclusions about the source of solutes. Weathering in silicate terrains can produce riverine water chemistry dominated by calcite and dolomite dissolution. This is especially true of high mountain areas, such as the Himalayas and the New Zealand Alps, where rapid erosion constantly exposes fresh traces of calcite to dissolution (Blum et al., 1998; Jacobson et al., 2001). Failure to correct for the calcite source of dissolved Ca in rivers draining silicate terrains has led to erroneous reasoning as to factors affecting global chemical weathering (e.g., Raymo, 1991; Huh et al., 1998).

Submarine Weathering of Basalt

The reaction of CO_2 dissolved in seawater with Ca and Mg silicate minerals in basalt can result in the formation of calcium and magnesium carbonates within the basalt (Staudigal et al., 1989). This is an additional process that could affect atmospheric CO_2, because oceanic and atmospheric CO_2 are in exchange contact. Some carbon cycle models include this process as a major control on CO_2 (Brady and Gislason, 1997; Sleep and Zahnle, 2001; Wallmann, 2001). The idea is simply that if atmospheric CO_2 rises, the concentration in the ocean rises, and there is enhanced uptake by submarine basalt weathering. In this way the process serves as a negative feedback for stabilizing atmospheric CO_2, and, in combination with silicate weathering on the continents, this would lead to greater dampening of CO_2 fluctuations over time. This process would have provided a more important feedback during the Precambrian (Sleep and Zahnle, 2001) because of a thermally enhanced mantle-crust-ocean-atmosphere carbon cycle in the early earth. Furthermore, Brady and Gislason (1997) have suggested that increases in deep sea temperature, due to the greenhouse effect of elevated CO_2, could lead to increased submarine basalt weathering and thereby act as an additional negative feedback.

The impact of the process of submarine weathering on atmospheric CO_2 over the Phanerozoic has been suggested to be minor to negligible (Berner, 1991; Berner and Kothavala, 2001). Reasons for this conclusion are as follows: (1) If the process is driven by changes in basaltic seafloor creation (spreading) rate, as is assumed in all models treating this phenomenon, then increased CO_2 uptake by faster seafloor spreading should be counterbalanced by faster CO_2 emission due to thermal degassing at mid-oceanic rises and subduction zones. (2) Caldeira (1995) demonstrated that basalt weathering to $CaCO_3$ cannot respond appreciably to changes in atmospheric and oceanic CO_2 because the pH of the interstitial water in contact with the basalt is buffered to a high value, and basalt dissolution is thereby greatly impeded. (3) Caldeira (1995) also showed that the thermal feedback mechanism of Brady and Gislason (1997) is untenable based on thermal gradients within oceanic crust. (4) Based on strontium isotope analyses, Alt and Teagle (1999) stated that 70–100% of the calcium in $CaCO_3$ found in altered submarine basalts is derived from seawater, not from the basaltic minerals. This means that most of the basalt is not actually being weathered but is simply providing a place for the interstitial precipitation of Ca^{++} and HCO_3^- from seawater. The neutral-to-alkaline interstitial seawater environment, due to hydrolysis of the basalt (Caldeira, 1995), may serve to raise pH and induce $CaCO_3$ precipitation. (5) Most Ca release from basalt is not due to dissolution by CO_2 (as carbonic acid) but by exchange with dissolved Mg in seawater (Alt and Teagle, 1999). (6) Assuming as a maximum that 30% of Ca is derived from basalt, and using the total formation rate of

$CaCO_3$ in submarine basalts given by Alt and Teagle, insertion of this degree of submarine weathering in the GEOCARB model (Berner and Kothavala, 2001) results in a very minor change in calculated CO_2 levels.

During the distant geologic past, submarine basalt weathering may have been more important than at present. In the present oceans more carbonate is deposited in deep sea sediments than in submarine basalts, and the carbonates later join in contributing to CO_2 degassing during subduction, but with basaltic carbonate being a lesser contributor. During the period before 150 Ma when there is little evidence for deep sea carbonate deposition (see chapters 3 and 4), it is possible that the subduction of basaltic carbonate could have been a more important factor in global degassing and in changes of atmospheric CO_2 level.

Summary

Much attention has been paid to the subject of silicate mineral weathering and how it relates to the evolution of atmospheric CO_2. As a result, the overall rate of uptake of atmospheric CO_2 over time is much better understood than the rate of release of CO_2 via degassing (see chapter 4). In the present chapter all of the major factors affecting silicate weathering have been examined. This includes (1) physical erosion as affected by mean continental relief; (2) temperature, as affected by the atmospheric greenhouse effect; solar evolution, and continental drift; (3) rainfall and runoff as affected by the greenhouse effect and continental drift; (4) land vegetation, as it evolved over time, and (5) lithology, especially the difference in weathering rates between carbonates and silicates and granites and basalts. Rate expressions for all these factors were derived in this chapter and presented in terms of nondimensional factors which can be applied to carbon fluxes involved in the long-term carbon cycle.

Although weathering is better understood than degassing, there are still some outstanding problems: (1) Is there a better method for quantifying the role of physical erosion over time as it has affected chemical weathering? (2) How have changes in the size and distribution of the continents affected the temperature of rocks undergoing weathering and the degree of flushing of these rocks by rainfall? (3) How did the rise of large vascular land plants affect weathering rate? There have been only a limited number of studies on this topic. (4) How did the rise of angiosperms affect weathering rate? This subject is poorly understood at present. (5) What is the quantitative effect of cosmic ray fluxes on climate and weathering? (6) How has the abundance of basaltic rocks exposed to weathering varied over time, and can the ratio of strontium isotopes in the oceans be used as a measure of basalt weathering? This all shows that there is still plenty to learn about silicate weathering and it how it affects atmospheric CO_2 over geologic time.

3

Processes of the Long-Term Carbon Cycle: Organic Matter and Carbonate Burial and Weathering

The organic subcycle of the long-term carbon cycle, where organic matter burial and weathering are involved, constitutes the major control on the evolution of atmospheric oxygen. It is also important as a secondary factor affecting atmospheric CO_2. Thus, it is important to better understand the processes whereby organic matter is buried in sediments and oxidized upon subsequent exposure to weathering during uplift onto the continents. This is especially true of the Paleozoic rise of land plants, which had a large effect on atmospheric CO_2 because of increased global organic burial due to the addition of plant debris to sediments. The burial of organic matter in marine sediments is impacted strongly by the availability of the nutrient elements, phosphorus and nitrogen, so a complete discussion of the cycling of organic carbon should involve some discussion of the cycles of these elements.

Carbonate burial is the ultimate sink for CO_2 derived from the atmosphere via the weathering of Ca and Mg silicates. The location of this burial, shallow water shelves versus the deep sea floor, is important because it affects the probability that the carbonate will be eventually thermally recycled and the carbon returned to the atmosphere. Carbonate weathering is the dominant process affecting river water composition and is a key component of the cycling of carbon. Its importance to the long-term carbon cycle is that, in order to calculate the removal of CO_2 from the atmosphere via Ca and Mg silicate weathering, it is necessary to correct total carbonate burial for that derived from carbonate weathering.

Organic Matter Burial in Sediments

At present, sedimentary organic matter burial occurs in swamps, lakes, reservoirs, estuaries, and in the open marine environment. The ultimate sources of the organics are land vegetation and marine phytoplankton. Also, soil organic matter, which is intimately associated with clay minerals, is eroded and transported to the sea by rivers (Hedges et al., 1994). A major question is how much of the total global burial is of marine or nonmarine origin. Recent work has shown that organic burial on land is much higher than previously recognized, especially as a result of human activities (Dean and Gorham, 1998; Stallard, 1998). In marine sediments, quantitatively separating marine from terrrestrial origin is difficult, but qualitative methods are available (e.g., Prahl et al., 1994). There is no doubt that more organic matter is carried to the sea than is buried there (Berner, 1982), which means that an appreciable proportion of terrestrially derived organics are destroyed in seawater (Hedges et al., 1997).

Within the oceans the locus of organic matter deposition and burial has been shown to be mainly in river deltas and on other areas of the continental shelves (Berner, 1982). Although sediments under areas of upwelling and high productivity are very rich in organic matter, the total burial in these areas is small compared to that in the organic-poor sediments of deltas and other near-shore depositional areas. Rates of sediment deposition in deltas, such as those of the Ganges/Brahmaputra and Amazon rivers, are so large that, even with low organic carbon concentrations, the global deltaic and shelf burial flux of carbon is more than 80% of all marine burial (table 3.1).

A major question is how the rates of global organic matter burial have varied in the geologic past. Knowledge of past global organic burial rates is essential for modeling the evolution of both atmospheric oxygen and carbon dioxide (see chapters 5 and 6). A key question is what were the factors that brought about increased or decreased organic matter sedimentation and preservation, the two steps leading to organic burial. For the nonmarine environment, there is no doubt that the rise and evolution of land plants brought about greater organic sedimentation (see next section). For the marine environment, there is some disagreement concerning the relative roles of sedimentation and preservation, both at present and in the past. There is no doubt that in the present ocean, more than 99% of organic matter sedimented to the sea floor becomes reoxidized, and not buried (Holland, 1978; Berner, 1989). However, it is the small fraction that becomes buried that controls O_2 production and CO_2 consumption, and this proportion could have varied in the past.

Table 3.1. Present (prehuman) burial fluxes of organic carbon in marine sediments.

Sediment type	Flux (10^{12} mol/year)
Deltaic-shelf sediments	4.4
Biogenous sediments underlying regions of high productivity	0.5
Shallow-water carbonate sediments	0.2
Pelagic sediments (not overlain by regions of high productivity)	0.2
Total	5.3

Values represent modern surface sedimentation rates and are corrected for 20% loss of carbon during early diagenesis and an anthropogenic contribution of about 50%. Data from Berner (1982).

Increased burial over geologic time could have been brought about by (1) increased global sedimentation of total solids (e.g., Berner and Canfield, 1989); (2) increased global biological productivity and increased organic sedimentation due to higher levels of dissolved phosphorus added to the oceans by weathering (e.g., Holland, 1994; Guidry and Mackenzie, 2000); (3) increased preservation due to development of anoxic bottom waters in a highly stratified ocean (Van Cappellen and Ingall, 1996); and (4) increased preservation due to a low oxygen content of the atmosphere and the oceans (Lasaga and Ohmoto, 2002). As with all other geologic phenomena, a combination of these factors could have been operative. For example, Van Cappellen and Ingall (1996) call for both increased preservation and increased productivity by the preferential release to seawater of phosphorus from sediments buried in anoxic bottom waters.

Organic production via photosynthesis in the present ocean is nutrient limited, and the principal nutrients are nitrogen and phosphorus, although iron may be limiting in some polar waters (Falkowski and Raven, 1997). Over geologic time there is disagreement whether phosphorus (e.g., Holland, 1978, 1994) or nitrogen (Falkowski, 1997) is the principal limiting nutrient. Settling this issue is important because the concentrations of the two elements in seawater are controlled by different processes. Nitrogen levels reflect a balance between nitrogen fixation, the conversion of atmospheric N_2 to nitrogen compounds accessible to organisms, and denitrification, the reduction of dissolved nitrate (the principal form of nitrogen in seawater) back to N_2. Weathering of rocks is unimportant in the nitrogen cycle. Phosphorus, in contrast, is minimally involved in atmospheric cycling, and the overall level of phosphorus in the ocean is a function of the riverine input from weathering of phosphate minerals and output by the burial of organic and inorganic phosphorus in sediments (Holland,

1978). An important additional factor is that photosynthesis can occur only in water shallow enough for the penetration of light. Falling dead organic debris is broken down to dissolved N and P via microbial respiration at depth and this, along with photosynthesis in shallow waters, leads to gradients in N and P with lower values at the surface and higher values at depth (Falkowski and Raven, 1997). Thus, vertical oceanic circulation, such as coastal upwelling, plays a role in supplying nutrients to shallow waters and fueling organic production and eventually burial in bottom sediments.

Because organic matter burial involves the production of oxygen and the consumption of CO_2, and marine organic burial is controlled largely by nutrient availability, considerable attention has been given to the cycles of phosphorus and nitrogen as to how they could have affected the burial of organic matter over geologic time (Holland, 1994; Van Cappellen and Ingall, 1996; Falkowski, 1997; Lenton and Watson, 2000; Wallmann, 2001). All of the P and N cycle hypotheses call for negative feedback mechanisms that limit fluctuations in the rates of organic burial. For Wallmann (2001), CO_2 consumption via organic burial is tied directly to the supply of phosphate to the oceans from the weathering of silicate, carbonate, and organic matter because phosphorus is associated with all three groups.

The various proposed nutrient feedback mechanisms can be represented in terms of a systems analysis diagram (figure 3.1). Positive responses are represented by plain arrows; negative responses are represented by arrows with bullseyes. Any loop with an odd number of bullseyes represents negative feedback or stabilization. Consider feedback loop A, one favored by Wallmann (2001) and Hansen and Wallmann (2003). Global warming due to elevated CO_2 brings about accelerated phosphate weathering and transport of P to the sea, leading to an increase in aqueous nutrient P. This in turn leads to greater organic carbon burial and greater CO_2 consumption, with the overall process producing negative feedback. The phosphorus-controlled feedback of Holland (1994), Colman and Holland (2000), and Van Cappellen and Ingall (1996) is shown by feedback loop B. An increase in atmospheric and oceanic O_2 leads to greater burial of phosphorus adsorbed on ferric oxides (FeP), which reduces the amount of aqueous nutrient P available for organic production. The lower concentration of nutrient P leads to less organic burial and a diminution in O_2 production, thus completing the feedback cycle. Then there is cycle C, favored by Falkowski (1997). Higher O_2 leads to diminished nitrogen fixation, leading to less organic matter production and burial and less O_2 production. Finally, there is an indirect interaction between atmospheric O_2 and CO_2. For example, a decrease in nutrient P in seawater, due to a rise in O_2, leads to decreased organic burial and higher CO_2.

There are periods of time during the Phanerozoic when it is believed that large portions of seawater became anoxic. Short-term (few million

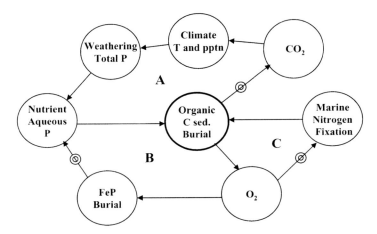

Figure 3.1. Systems analysis diagram for the effects of nutrients on O_2, CO_2, and the organic subcycle. Arrows with bullseyes represent negative response; arrows without bullseyes represent positive response. A complete cycle (loop) with an odd numbers of bullseyes means negative feedback and stabilization; a complete cycle with an even number of bullseyes, or no bullseyes, means positive feedback and (normally) destabilization. The three loops shown are all for negative feedback and stabilization of the atmospheric gases.

year) anoxic ocean events (AOEs) during the Mesozoic have been well documented (Arthur et al., 1985; Jenkyns, 1988; Arthur and Sageman, 1994). A much longer anoxic period occurred during the early Paleozoic, when the black shales of the graptolite facies were deposited (Berry and Wilde, 1978; Wilde, 1987) and the average ratio of organic carbon to pyrite sulfur buried in sediments globally was unusually low (Berner and Raiswell, 1983). This is shown in figure 3.2. Low organic C/pyrite S is characteristic of deposition in anoxic bottom waters—in other words, in euxinic environments characterized by the pyrite-rich present day deep Black Sea sediments (Berner and Raiswell, 1983). (The prominent maximum in C/S of figure 3.2 is discussed below in "Land Plant Evolution.") If euxinic conditions lead to greater preservation of organic matter, then increased global anoxicity could have led to greater organic burial rates during these anoxic periods. Greater burial during the AOEs is suggested by increases in $\delta^{13}C$ of seawater indicating greater removal of light carbon in organic matter (e.g., Arthur et al., 1985). Also, there is a good correlation between high global organic burial rate and times of formation of oil-source rocks (Berner, 2003).

There is an argument concerning whether anoxic waters preserve organic matter better than oxic waters. Pederson and Calvert (1990) have claimed that greater burial of organic matter is due more to higher pro-

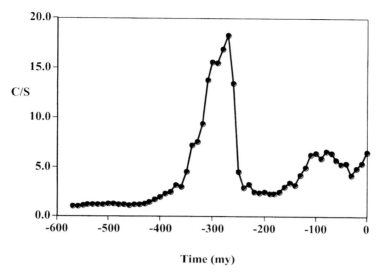

Figure 3.2. Plot of the rate of burial of organic carbon divided by the rate of burial of pyrite sulfur (C/S) over Phanerozoic time. Low C/S ratios are characteristic of organic burial in euxinic (anoxic bottom water) basins, high values in terrestrial fresh water swamps, and intermediate values in normal marine (noneuxinic) sediments (Berner and Raiswell, 1983). The very low values during the early Paleozoic (550–400 Ma) suggest appreciable burial in global-scale anoxic bottom waters. The large maximum centered around 300 Ma represents a shift of major deposition from the marine to the terrestrial environment due to the rise of vascular land plants. Deposition on land results in high C/S because of the much lower sulfur content of fresh water compared to seawater.

ductivity, with a consequent higher flux of organics to the bottom, than to better preservation in anoxic bottom waters. However, this idea is opposed by many authors. For example, Canfield (1994) has shown that the fraction of sedimenting organic matter buried and preserved is usually higher in anoxic as compared to oxic bottom waters. Hedges et al. (1999) have further shown that organic matter preservation is related inversely to the time of exposure of the organics to dissolved O_2 in the marine environment. This links the two hypotheses of sedimentation rate and anoxic preservation into a combined explanation for organic matter burial.

The total rate of burial of organic carbon over the Phanerozoic, both marine and nonmarine, has been calculated via two independent methods. (Details of these methods are presented in chapter 6.) In the first, the abundance of different rock types and their organic carbon contents, as a function of time (Ronov, 1976), has been used for direct calculation (Berner and Canfield, 1989). The other method involves the use of stable carbon isotopes ^{13}C and ^{12}C. During photosynthesis there is a large fractionation such that the produced organic matter is enriched in ^{12}C. To express this change, one can use the simple definition:

For land plants: $\Delta^{13}C = \delta^{13}C(plant) - \delta^{13}C$ (atmospheric CO_2)   (3.1)

For marine phytoplankton: $\Delta^{13}C = \delta^{13}C(phytoplankton) - \delta^{13}C(DIC)$

where

$\delta^{13}C = [^{13}C/^{12}C)/(^{13}C/^{12}C)stnd - 1]1000$
DIC = total dissolved inorganic carbon in seawater

The range of $\delta^{13}C$ values for terrestrial plants, soil organic matter, and marine plankton is about −20 to −30‰ (this excludes C-4 land plants that evolved near the end of the Phanerozoic).

Values of $\delta^{13}C$ for dissolved inorganic carbon in the oceans over time are approximated by the $\delta^{13}C$ of calcareous fossils and limestones (e.g., Veizer et al., 1999). A plot for carbonate fossils is shown in figure 3.3. Assuming isotopic exchange equilibrium between the oceans and the atmosphere, the value of $\delta^{13}C$ for atmospheric CO_2 should be about 7‰ lower than the oceanic DIC value. Thus, any change in the $\delta^{13}C$ value of the oceans should be mirrored by a similar change in that for atmospheric CO_2.

The data of figure 3.3 show a broad maximum of $\delta^{13}C$ centered around the Permian-Carboniferous boundary at 290 Ma. This maximum is most readily interpreted as the result of enhanced removal of light carbon from seawater and the atmosphere due to the burial at that time of greater amounts of isotopically light organic matter (changes in the burial of limestone have a much lesser effect on oceanic $\delta^{13}C$). These results are in agreement with the results of measurements of the organic carbon contents of sediments of different ages mentioned above. In other words, there was more organic carbon burial in Permo-Carboniferous sediments. Much of the additional Permo-Carboniferous organic carbon burial was due to the rise of large vascular land plants (see next section).

The fractionation of carbon isotopes $\Delta^{13}C$ (equation 3.1) during photosynthesis has likely changed over Phanerozoic time. This is suggested by the compilation of $\delta^{13}C$ data for Phanerozoic sedimentary carbonates and organic matter by Hayes et al. (1999; see figure 3.4). Hayes et al. ascribes the drop in $\Delta^{13}C$ during the late Cenozoic mainly to a drop in atmospheric CO_2. Marine photosynthesis is affected by changes in dissolved CO_2 (Arthur et al., 1985: Jasper and Hayes, 1990; Freeman and Hayes, 1992) and terrestrial photosynthesis by changes in O_2 (Beerling et al., 2002), and these dependencies can be used to deduce past levels of these gases in the atmosphere (see chapters 5 and 6).

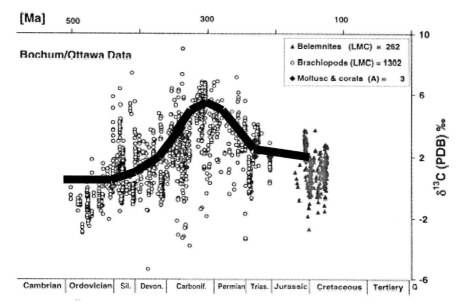

Figure 3.3. Plot of $\delta^{13}C$ versus time for the Ordovician to the Cretaceous Periods. The data are fitted with a "mean" value curve that is biased somewhat toward higher $\delta^{13}C$ values because of the distinct possibility that lower values are too low because of diagenetic alteration of the original marine isotopic signal. This fitted curve is used in GEOCARB modeling. (From Berner, 2001, after Veizer et al., 1999.) © 2001 with permission from Elsevier.

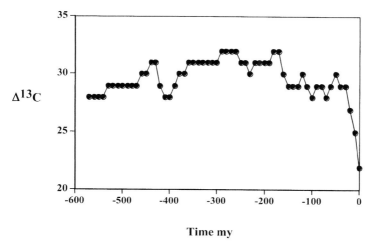

Figure 3.4. Plot of $\Delta^{13}C$, the difference between $\delta^{13}C$ of coexisting calcium carbonate and organic matter for the Phanerozoic. (Data from Hayes et al., 1999.)

Land Plant Evolution

The rise of large vascular land plants during the Devonian perturbed the long-term carbon cycle not only by accelerating the weathering of silicate rocks, but also by bringing about the removal of atmospheric CO_2 and production of O_2 by increased burial of organic matter in sediments. The increased burial was due primarily to the creation by the plants of a new microbially resistant form of organic matter, lignin. Lignin is unique to woody land plants, and it is decomposed only by a very limited biota, such as white rot fungi (Hedges et al., 1988; Hedges and Oades, 1997). Ligniferous plant debris was deposited in terrestrial coal swamps, but also in the marine environment after transport there by rivers, where it joined the burial of marine-derived organics. Evidence that large amounts of lignin were buried in the Carboniferous and Permian is indicated by the vast coal reserves of this time span. Coal is formed primarily during deep burial by the transformation of lignin to more carbon-rich aromatic substances (Thomas, 2002).

There is some suggestion that initially lignin burial was unusually large because of the relative absence of the appropriate decomposing fungi (Robinson, 1990, 1991), which evolved somewhat later in the Paleozoic. This could account for the fact that combined Permian and Carboniferous coal reserves outweigh those for all other periods (Bestougeff, 1980), even though these coals are the oldest and have been subjected to loss by erosion over the longest time. This is illustrated in table 3.2. A uniquely high rate of burial of organic matter during the Permo-Carboniferous is also shown by the relative abundance of coal basin sediments at that time relative to subsequent periods (figure 3.5; Ronov, 1976).

Table 3.2. Abundance of potentially recoverable coal as a function of geological age.

Period	Coal mass (10^{15} mol/my)
Cambrian	0
Ordovician	0
Silurian	0
Devonian	1
Carboniferous	31
Permian	53
Triassic	1
Jurassic	30
Cretaceous	15
Tertiary	19

Masses are normalized per unit time to correct for different lengths of the geological periods. Data from Bestiougeff (1980) and Raiswell and Berner (1983).

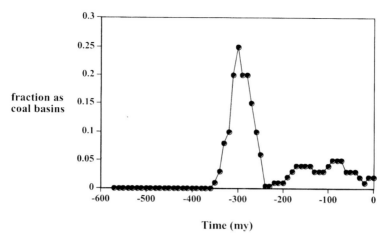

Figure 3.5. Abundance of coal basin sediments as a function of time expressed as a fraction of total terrigenous sediments (sandstones and shales). (Data from Ronov, 1976.)

Another factor contributing to the high abundance of Permian and Carboniferous organic matter deposition is the paleogeography of that time (Stanley, 1999). There was an abundance of flat land on the supercontinent of Pangaea, especially as coastal lowlands. Because sea level was relatively low at the time, areas that would ordinarily be covered by epeiric seas had become low-lying land, and, combined with high rainfall and low relief, very shallow groundwater levels gave rise to abundant freshwater and brackish water swamps. Preservation of deposited organic matter on land is appreciable only in O_2-free environments, and swamps provided a place for this to happen. Little organic matter in normally well-drained soils is preserved and buried into the geologic record because of its oxidation to CO_2 by atmospheric oxygen. Therefore, most paleosols are very low in organic matter (Retallack, 1990).

Burial of organic matter in swamps is accompanied by limited pyrite (FeS_2) formation, resulting in a high ratio of organic carbon to pyrite sulfur in freshwater sediments as compared to marine sediments. This is because there is so much less dissolved sulfate, the source of pyrite sulfur, in fresh water than in seawater (Berner and Raiswell, 1983). As a result, the high rate of burial of nonmarine organic matter during the Permo-Carboniferous is accompanied by a high C/S ratio (figure 3.2).

There is no clear evidence that the rise of angiosperms during the Cretaceous resulted in greater global rates of organic matter burial. Cretaceous and Cenozoic coal is not markedly more abundant compared to coal deposited earlier. A notable anomaly is the Triassic, when coal deposition (table 3.2; figure 3.5), and organic matter burial in general (Berner and Canfield, 1989), was extremely low. This may have been due to a lack of proper climatic and topographic conditions for the formation of coal swamps on Pangaea, as demonstrated by the paucity of Triassic coal basin sediments (figure 3.5). The rise of C-4 plants during the late Miocene (Cerling et al., 1997), as typified by warm-climate grasslands, is an additional factor to be considered in organic matter burial. The C-4 plants fractionate carbon isotopes less than most other (C-3) plants (δ^{13}C averages around −15‰), and contribution of C-4 plant debris to total global organic matter burial could have played a role in the latest portion of the observed drop in the mean value of Δ^{13}C (Hayes et al., 1999) during the Cenozoic (figure 3.4).

Weathering of Organic Matter

Much less attention has been paid to the weathering (oxidation) of organic matter than to its burial. This is a shame because both processes equally affect the level of atmospheric CO_2 and O_2. Recently, experimental evidence has been presented that shows that the oxidation of coal (as a representative of sedimentary organic matter) in water results in the formation of oxygen-containing organic compounds before complete oxidation to CO_2 (Chang and Berner, 1999). The kinetics of the coal–oxygen reaction deduced by Chang and Berner have been applied recently to a model for global organic matter oxidation (Lasaga and Ohmoto, 2002). However, the Lasaga and Ohmoto model overly simplifies organic weathering in that it does not take into consideration several complicating factors: (1) Organic matter in sedimentary rocks (kerogen) is dispersed in shales with complicated geometries, including adsorption within clay minerals (Hedges and Keil, 1995). Lasaga and Ohmoto treat organics as grains of coal as in the Chang experiments. (2) Oxygen enters shales by diffusion down concentration gradients, both in air and in interstitial solution, not by being in a uniformly well-mixed "soil" with atmospheric O_2 composition that was assumed by Lasaga and Ohmoto. (3) Kerogen oxidation in shales is mediated and probably accelerated by bacteria (Petsch et al., 2001). Thus, use of the Chang oxidation rates by Lasaga and Ohmoto, which were obtained via abiotic experiments, may be incorrect.

Oxidation of sedimentary organic matter in a variety of shales of different ages has been studied in the field (Petsch et al., 2000). Petsch et al. demonstrated gradients of organic carbon in shales exposed to weathering and found that organic N and organic S are lost at the same rate as organic C. An example of their results from one area is shown in figure 3.6. They also found that the organic matter picks up O_2 to form oxygen-containing functional groups, as also found in the Chang experiments, and that the decomposition processes are aided by bacterial activity (Petsch et al., 2001). Finding nonzero organic carbon at the exposed surface of the shale (Petsch et al., 2000) at some localities indicates that rates of erosion can exceed rates of oxidation, resulting in the transport and burial of old, unoxidized organic matter in modern sediments. Burial

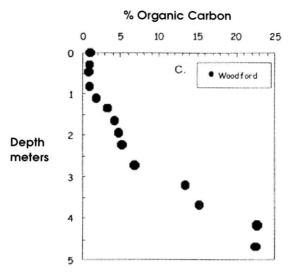

Figure 3.6. Plot of organic carbon versus depth for a single stratum in the Upper Devonian Woodford shale, from the Arbuckle Mountains, Oklahoma, USA. Because sampling was restricted to the same depositional layer, loss of carbon toward the surface is due to oxidation during weathering and not to change in organic content at the time of burial. (Data from Petsch et al., 2000.)

of this recycled, old organic matter has been demonstrated by studies of modern sediments (e.g., Eglinton et al., 1997; Blair et al., 2003), and it complicates modeling of the long-term carbon cycle.

A theoretical model for organic matter weathering in shales is being constructed (Bolton et al., unpublished ms) that incorporates erosion, organic matter oxidation, O_2 diffusion, shale porosity and permeability, organic matter and shale texture, pyrite oxidation (most organic-rich marine shales contain abundant pyrite), and measured gradients of organic C and pyrite S. The fundamental question being addressed is whether the rate of oxidative weathering of both organic matter and pyrite is controlled by O_2 transport into the shale or by the exposure of the organic matter and pyrite to the atmosphere by erosion. If the former is true, then the rate of weathering becomes a function of the level of O_2 in the atmosphere. It has been assumed by some authors (e.g., Holland, 1978; Berner, 2001) that organic weathering is independent of atmospheric O_2 and dependent only on uplift and erosion. Preliminary results of the modeling suggest that this is correct, but further work is needed.

With regard to the factors affecting the oxidative weathering of organic matter, all of those that affect silicate weathering, as discussed in chapter 2, need to be, at least, considered. Certainly uplift and erosion is important in exposing organic matter to the atmosphere. Bacterial respiration is strongly temperature dependent, and if organic matter oxidation is mainly bacterially mediated, feedback factors based on temperature change need to be considered. Runoff, as affected by continental drift, should be included if organic weathering is not limited only by uplift and erosion. More flushing of the organic matter by O_2-containing water should lead to faster weathering. The remaining weathering-impacting processes—the effect of plants and lithology—probably play lesser roles in organic oxidation. Thus, in modeling the long-term carbon cycle, an expression for organic matter weathering rate needs to include the modifying parameters $f_R(t)$ for erosion (equation 2.1 or 2.5), $f_{AD}(t)$ for runoff (equation 2.27), and $f_{Bt}(t)$ for temperature as it relates to CO_2, solar evolution, and land temperature (equation 2.29). Finally, it remains uncertain whether the O_2 level of the atmosphere should be also included in a weathering rate expression for organic matter, as done by Lasaga and Ohmoto (2002). A final decision awaits the results of better models for organic matter weathering and more field measurements on shales.

Carbonate Weathering

The most notable aspect of present-day carbonate weathering is that it is so much faster than silicate weathering. Silicate terrains, such as portions of the high Himalayas and the New Zealand Alps (Blum et al., 1998; Jacobson et al., 2003) that contain only traces of carbonates are marked by water chemistries dominated by carbonate dissolution. On a much grander scale, although silicates cover most of the surface area of the continents, the global average chemical composition of river water is characteristic of that expected for carbonate weathering (Meybeck, 1987; Berner and Berner, 1996; Gaillardet et al., 1999). In other words, limestone weathering dominates global water chemistry.

The factors affecting the weathering of Ca and Mg carbonates (calcite and dolomite) are similar to those affecting silicate weathering. This includes temperature, as it is affected by the atmospheric greenhouse effect, solar radiation, and continental drift; hydrological changes accompanying changes in global temperature and paleogeography; and the rise and evolution of land plants. However, relief is not as important as it is for silicate weathering. This is because limestones undergo subsurface dissolution by groundwaters regardless of relief. An outstanding example of the development of subsurface limestone dissolution on flat ground is shown by the common sink holes of southern Florida, USA, which can be easily observed from the air by anyone flying into the Miami airport. Also, the flow of water through limestones (and dolostones) is increased as more and more dissolution, including cave formation, proceeds. Silicates require access of water to primary minerals, which can be impeded by a thick, relatively impermeable mantle of clay weathering products, as in the Amazonian lowlands. Limestone weathering, in contrast, does not involve the production of clays because there is only simple dissolution of the primary minerals.

As pointed out in the Introduction, on a multimillion-year time scale, carbonate weathering has essentially no direct effect on atmospheric CO_2. However, it has important indirect effects. First, greater carbonate weathering, for a steady-state ocean, means greater carbonate burial, and greater carbonate burial, if in deep sea sediments, means greater carbonate subduction with greater CO_2 degassing. If increased carbonate weathering is forced by higher atmospheric CO_2, this leads to positive feedback (Hansen and Wallmann, 2003). Second, in calculating the rate of CO_2 uptake by past Ca and Mg silicate weathering, it is necessary to also know the rate of carbonate weathering. Total carbonate buried in sediments comes from the weathering of both carbonates and Ca-Mg silicates. Therefore, to calculate the flux of CO_2 taken up by the conversion of Ca-Mg silicates to carbonates (reaction 1.5), one must subtract from total carbonate burial that derived from carbonate weathering (equation 1.13).

Models of the long-term carbon cycle (e.g., Berner and Kothavala, 2001; Wallmann, 2001) include formulation of an expression for carbonate weathering that is similar to that used for silicate weathering. However, in GEOCARB modeling there are some changes from the expression used for silicate weathering: (1) A separate temperature/CO_2 feedback factor $f_{Bc}(CO_2)$ is used that differs from $f_{Bt}(CO_2)$ because of a different temperature coefficient (Z) for carbonate weathering (Drake and Wigley, 1975); (2) the paleogeographic runoff/weathering term $f_{AD}(t)$ is not raised to the power 0.65 because carbonates dissolve so fast that they saturate the waters with which they come into contact, and there is no dilution at high runoff (Drake and Wigley, 1975; Stallard, 1995); and (3) the parameter $f_R(t)$ is not applied to limestone weathering because of the lesser role of relief, as discussed above.

An additional parameter is added to the expression for carbonate weathering based on the results of Bluth and Kump (1991) for the proportions of land areas underlain by carbonates over the Phanerozoic. Results are derived from a series of global paleolithologic maps laboriously compiled from large amounts of the data of Ronov et al. (1984, 1989). From the results of Bluth and Kump (1991), the dimensionless parameter $f_{LA}(t)$ is derived (Berner, 1994; Wallmann, 2001) that is applied to carbonate weathering:

$$f_{LA}(t) = f_L(t)\, f_A(t) \qquad (3.2)$$

where

    $f_L(t)$ = fraction of total land area covered by carbonates/the same fraction at present

    $f_A(t)$ = total land area/total present land area.

In the model of Hansen and Wallmann (2003), changes in carbonate weathering fluxes, as a result of changes in the exposure area of carbonates on the continents, are used to drive changes in organic matter burial because the phosphorus released during the weathering of carbonate sediments stimulates organic production upon transfer to the oceans (see figure 3.1).

Carbonate Deposition and Burial

In steady-state modeling of the long-term carbon cycle, the global rate of burial of carbonates F_{bc}, is simply a dependent variable calculated from the rate of input of carbon to the oceans plus atmosphere from weathering and CO_2 degassing. F_{bc} is calculated from the combination of equations (1.10), (1.11), (1.12), and (1.13) and is a consequence of the necessity of having global C inputs balanced by outputs. The only variation in F_{bc}, independent of variations in the inputs, is due to variations in $\delta^{13}C$, which governs the relative proportions of carbonate versus organic matter burial.

In non–steady-state models, such as those of Berner et al. (1983) and Wallmann (2001), global carbonate burial is not assumed to be simply dependent on inputs to the sea, but it is driven by the degree of supersaturation of the ocean. The formulation used in these studies is:

$$F_{bc} = k(C_{Ca}C_{HCO_3}{}^2 - K_{sp} P_{CO_2})$$ (3.3)

where C is concentration, P is partial pressure, k is a rate constant, and K_{sp} is the mean equilibrium constant in seawater for the reaction:

$$CO_2(gas) + CaCO_3 + H_2O \leftrightarrow Ca^{++} + HCO_3^-$$ (3.4)

At equilibrium, $C_{Ca}C_{HCO_3}{}^2 / P_{CO_2} = K_{sp}$ and $F_{bc} = 0$. Equation (3.3) ensures stability of oceanic composition by providing negative feedback against the buildup of excessive levels of dissolved calcium and bicarbonate in seawater. Also, at reduced levels of Ca and HCO_3^-, equation (3.3) shows that dissolution of already deposited calcium carbonate should occur.

There are some problems with this simple rate law. First, the ocean is not a uniform reservoir, and it is divided into shallow waters where $CaCO_3$ is supersaturated, and deep waters where it is undersaturated. Thus, precipitation occurs in shallow waters and dissolution occurs in deeper waters. Second, the simple rate law does not recognize the fact that most carbonate precipitation from seawater takes place biologically (Morse and Mackenzie, 1990). Corals, molluscs, calcareous algae, forams, and so on, secrete calcite and aragonite by biochemical processes that do not simply follow the rate laws formulated for inorganic precipitation. Third, dissolution in deep sea sediments is complicated by reactions above, at, and below the sediment–water interface, and it can even occur in sediments overlain by supersaturated seawater. In the latter case this is due to an excess of carbonic acid within the sediments caused by the oxidative decay of organic matter (Emerson and Archer, 1990). Regardless of all these problems, however, as a first approximation it can be assumed that, on average, the whole ocean can be treated by simple kinetics for both precipitation and dissolution as shown above. An even simpler model that is much easier to use and is almost equivalent in results to equation (3.3), is to assume that the oceans are saturated at all times with calcium carbonate (Caldeira and Berner, 1999).

In contrast to total burial rate, the locale of burial of carbonates is an independent variable that is important to the long-term carbon cycle. Calcium carbonate is deposited in the ocean as aragonite, calcite and highly magnesian calcite in shallow waters and as calcite in deep waters and the shallow water carbonates are sometimes converted to dolomite during early or late diagenesis (Morse and Mackenzie, 1990; Arvidson and Mackenzie, 1999). Carbonate deposited in the deep sea is much more likely to undergo subduction at plate boundaries and thermal decomposition at depth than the shallow water carbonates. Thus, any shift of deposition between shallow platforms and the deep sea could lead to changes in global rates of CO_2 degassing and to changes in atmospheric CO_2 level. This is believed to have happened in the geological past.

The principal Mesozoic and Cenozoic sources of deep sea or pelagic carbonate are the skeletal remains of calcareous plankton, chiefly foraminifera and coccoliths. These organisms did not arise until the Jurassic period, about 150 million years ago. Since that time they have expanded and grown in abundance. The remains of the calcareous plankton fall to the deep sea floor and accumulate there to form carbonate-rich pelagic sediment. Partly because of the evolution of the calcareous plankton, pelagic deposition has increased over the past 150 million years so that at present more than half of global carbonate sedimentation is in the deep sea (Milliman, 1974). Added to this is the loss of shallow water shelves and platforms available for carbonate deposition. Walker et al. (2002) have calculated that the rate of shoal water carbonate deposition has decreased sixfold over the past 70 million years because of a loss of sufficiently warm, shallow water areas. To avoid accumulation of dissolved carbon in the ocean, this has meant a corresponding increase in deep sea deposition.

Before 150 Ma there is little direct evidence for pelagic carbonates, even though there is strong evidence of a decrease in shallow-water carbonates with decreasing age over the entire Phanerozoic (Walker et al., 2002). This decrease, shown in figure 3.7, is unexpected because older rocks should be preferentially lost by longer periods of exposure to erosion. The lack of younger shallow-water carbonates has been

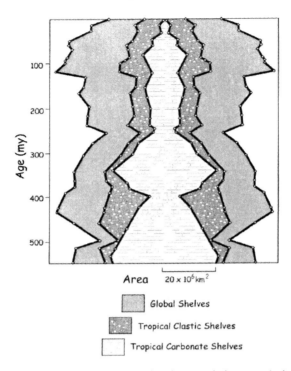

Figure 3.7. Relative abundance of tropical carbonate shelves, tropical clastic shelves, and total global shelves versus time for the Phanerozoic. Note the consistent loss of carbonate shelves with time. (After Walker et al., 2002.) © 2002 by the University of Chicago.

explained by the loss of warm-water shelves with time (Walker et al., 2002). Loss of shallow-water areas available for carbonate deposition should be accompanied by an increase in deep water deposition. However, there is no evidence for mid-ocean pelagic carbonates older than 150 Ma, indicated by an absence of chalks and carbonate sediment-containing ophiolites older than this time (Boss and Wilkinson, 1991). (Ophiolites are samples of ancient mid-ocean ridge crests.) Thus, there is a quandary as to where carbonate deposition occurred before 150 Ma. This is especially true of the period 350–250 Ma (Walker et al., 2002) when the abundance of carbonate shelves decreased by a factor of about 4 (figure 3.7).

Summary

Organic matter burial is a major process for transferring carbon dioxide from the surficial system to the rock record. It is also the principal process for the production of atmospheric oxygen. Rates of burial have been calculated from the abundance of organic matter in sedimentary rocks as a function of time and through the use of the carbon isotopic composition of seawater as recorded by sedimentary carbonates. Both methods demonstrate that organic burial increased appreciably during the Permian and Carboniferous due primarily to the rise of large land plants containing microbially resistant lignin, and secondarily to the abundance of coastal lowlands appropriate for the formation of coal basins. During periods when the oceans were low in dissolved oxygen, there was probably greater burial of organic matter, although the role of bottom-water anoxicity in enhancing organic matter preservation has been challenged by some workers.

The weathering of organic matter is as important to the long-term carbon cycle as its burial, but there has been little study of weathering. Organic weathering is a major sink for O_2 and a major source for CO_2. A critical question for carbon cycle modeling is whether organic weathering responds to changes in atmospheric O_2. It has been assumed by most workers that weathering is O_2 independent and controlled by the rate of uplift of the organic matter into the zone of oxidation. However, this is not a settled question, and research is needed to try to answer the question of rate control. This is important because if organic weathering is O_2 dependent, the weathering process acts as a negative feedback control on excessive O_2 excursions.

Carbonate weathering impacts the long-term carbon cycle, not as a direct control on atmospheric CO_2, but as a necessary flux term for calculating the level of atmospheric CO_2 consumed by silicate weathering. Rate expressions for carbonate weathering have been constructed that are similar to those for silicate weathering. The rate of global carbonate deposition can be treated as a dependent variable and can be calculated from global carbonate and silicate weathering plus imbalance in the organic carbon subcycle. However, the location of carbonate deposition is critical as to the liklihood of the carbonate being decomposed later by thermal processes leading to the degassing of CO_2 to the ocean and atmosphere. There is a higher probability of thermal decomposition for deep sea pelagic carbonate because it is transferred to great depth in subduction zones, whereas shallow-water carbonates are not. The history of shallow versus deep water carbonate deposition is thus important to deducing the history of CO_2 degassing.

4

Processes of the Long-Term Carbon Cycle: Degassing of Carbon Dioxide and Methane

Degassing of CO_2 and CH_4 to the atmosphere and oceans is the process whereby carbon is restored to the surficial system after being buried in rocks. Carbon dioxide is released by a variety of processes. This includes volcanic emissions from the mantle and metamorphic and diagenetic decarbonation of limestones and organic matter. Volcanic degassing can occur over subduction zones, at mid-ocean rises, on the continents, and in the interior of oceanic plates. Degassing can be sudden and violent, as during volcanic eruptions, or slow and semi-continuous in the form of fumaroles, springs, gas vents, and continually degassing volcanic vents. An outstanding example of the latter is Mt. Etna, which contributes about 10% to total global degassing (Caldeira and Rampino, 1992). Metamorphic degassing is concentrated in zones of seafloor subduction (Barnes et al., 1978), crustal convergence (Kerrick and Caldeira, 1998), and crustal extension (Kerrick et al., 1995). Most methane degassing on a geologic time scale occurs from organic matter diagenesis slowly from coal, oil, and kerogen maturation and suddenly from methane hydrate breakdown. A smaller amount of CH_4 emanates from mid-ocean hydrothermal vents.

Degassing Rate of CO_2

Estimates of present-day global volcanic degassing rates are under constant revision (e.g., see Gerlach, 1991; Brantley and Koepenick, 1995;

Sano and Williams, 1996; Marty and Tolstikhin, 1998; Kerrick, 2001). A compilation of recent estimated rates of most degassing processes is shown in table 4.1. A constraint on estimates is that none can exceed total global degassing. The latter can be determined from the steady-state assumption that CO_2 release by global degassing must be balanced by global uptake by Ca and Mg silicate weathering (Berner, 1990; Berner and Caldeira, 1997). (This assumes essential balance of the organic C subcycle.) Global Ca and Mg silicate weathering, based on river fluxes of these elements to the sea, has been estimated to be about $6 \pm 3 \times 10^{18}$ mol/my (Berner, 1990). Gaillardet et al. (1999) estimate a minimum value for Ca and Mg silicate weathering of 3.6×10^{18} mol/my. The individual estimates shown in table 4.1, being less than 6×10^{18} mol/my, are thereby acceptable, and one can derive an estimated total degassing from these data that essentially agrees with the range derived from the rate of CO_2 uptake by weathering.

The global rate of carbon degassing and how it has changed for all of geologic time has been modeled by Tajika and Matsui (1992), Sleep and Zahnle (2001), and Franck et al. (2002). A general result of these models is that over time the carbon content of the mantle has changed, but the rate of change during the past 500 million years has been small. This suggests that net loss of carbon to the mantle or gain from the mantle over the Phanerozoic has been sufficiently small that the mass of crustal carbon can be assumed to have remained essentially constant (Tajika and Matsui, 1992). This greatly simplifies Phanerozoic carbon cycle modeling. However, the rise of planktonic organisms in the Jurassic and the subsequent transfer of carbonate deposition from shelves and platforms to the deep sea probably has resulted in the addition of extra carbon to the mantle by seafloor subduction over the past 150 million years. This is taken into consideration in GEOCARB modeling (Berner, 1994; Berner and Kothavala, 2001; see chapter 3) and models of Wallmann (2001). Before this time, however, it is as-

Table 4.1. Present-day global degassing fluxes for CO_2.

Source	Flux (10^{18} mol/my)	Reference[a]
Release from mantle at mid-ocean rises (MOR)	1–3	1
Release in arc volcanoes from subducted $CaCO_3$	2–3	1, 2
Release in arc volcanoes from mantle	0.3–0.5	1, 2
Release from mantle in intraplate volcanoes (plumes)	0.5–3	1, 3
Release from *all* subaerial volcanoes	2–2.5	4
Release during metamorphism of continental carbonates	1	5
Total global degassing	4–10	

References: 1, Marty and Tolstikhin (1998); 2, Sano and Williams (1996); 3, Wallmann (2001); 4, Kerrick (2001); 5, Kerrick et al. (1995).

sumed that the mass of crustal carbon remained constant during the remainder of the Phanerozoic.

In many carbon cycle models (e.g., Berner et al., 1983; Fancois and Godderis, 1998; Tajika, 1998; Berner and Kothavala, 2001; Wallmann, 2001; Francois et al., 2002), changes in rates of Phanerozoic volcanic and metamorphic global degassing have been assumed to be directly proportional to rates of seafloor area creation, in other words spreading rate. For example, Tajika (1998) separates degassing into four types: (1) mid-ocean ridge volcanism, (2) hot-spot volcanism, (3) metamorphism of carbonate at subduction zones, and (4) metamorphism of organic carbon at subduction zones. All but hot-spot volcanism are assumed to be driven by changes in seafloor spreading rate. A similar approach is used by Wallmann (2001). How spreading rate affects global CO_2 degassing has been expressed in GEOCARB modeling (e.g., Berner, 1991) by the dimensionless parameter $f_{SR}(t)$:

$$f_{SR}(t) = \text{seafloor spreading rate (t)/seafloor spreading rate (0)} \qquad (4.1)$$

where (0) represents the present. Spreading rate dependency makes sense for the degassing that occurs at spreading centers and at subduction zones, but it also has been extended to be a measure of total tectonically induced degassing, whether on land or at sea (Berner et al., 1983; Berner, 1991). In this case the parameter $f_{SR}(t)$ is assumed to be equal to $f_G(t)$ to represent all degassing. This may be incorrect because spreading rates are not a good measure of CO_2 degassing resulting from magmatism and metamorphism accompanying continent–continent and arc–continent collisions (Kerrick and Caldeira, 1998).

Another source of CO_2 that is probably not a function of seafloor spreading rate is that resulting from mid-plate volcanism. During the Cretaceous period large mid-plate volcanic plateaus arising from mantle "superplumes" were built on the Pacific seafloor (Larson, 1991). This could have led to extra CO_2 emission to the atmosphere and has been called upon by many workers to explain Cretaceous warming and high CO_2 levels (Larson, 1991; Kaiho and Saito, 1994; Tajika, 1999; Wallmann, 2001). Kaiho and Saito correlate mid-plate volcanism, along with ridge and back-arc volcanism, with the temperature of Cretaceous seawater as recorded by oxygen isotopes. However, Heller et al. (1996) have challenged the quantitative significance of this "extra" volcanism on geophysical grounds.

Kerrick (2001) has pointed out additional problems with using spreading rate as a measure of global degassing: (1) In the circum-Pacific arc, there is no correlation between the number of volcanoes and subduction rate; (2) as oceanic basalts age, they undergo an increase in carbonate content by submarine weathering so that older subducting crust would release more CO_2 than younger crust; (3) the rise of calcareous plankton in the Mesozoic caused an increase in the carbonate content of subducting sediments; and (4) the extent of subduction-zone metamorphic decarbonation depends on pressure and temperature gradients as they affect the stability of carbonate minerals. In a global and long-term context one can answer these additional criticisms. First, there is an overall qualitative correlation between subduction and metamorphic degassing as shown by a global map of the distribution of CO_2-rich fumeroles and soda springs by Barnes et al. (1978). Volcanic density is not the sole measure of total degassing at any one place. Second, the average age and average carbonate content of subducting basalt should vary inversely with global average spreading rate. Third, the role of calcareous plankton in transferring carbonate deposition to the deep-sea floor is included in most carbon cycle models (Berner, 1991, 1994; Berner and Kothavala, 2001; Wallmann, 2001; see also chapter 3). Finally, much degassing at subduction zones is probably volcanic, not metamorphic, in origin, so the suggestion that metamorphic decarbonation varies inversely with subduction rate (Kerrick et al., 2003) applies only to a portion of arc degassing.

Following the earlier work of Budyko and Ronov (1979), an alternative to the use of seafloor spreading rate as a proxy for CO_2 degassing was offered by Kerrick et al. (2003). They assume that the CO_2 flux from volcanic arcs can be correlated with the rate of andesitic extrusion and that the rate of andesitic extrusion through time could serve as a measure of arc degassing. Examining modern volcanic data, they conclude that the present degassing rate per unit volume of andesite is 3.5×10^8 mol/my/km^3. From this value and a volumetric estimate of Late Cretaceous andesites, they conclude that CO_2 emission rates for the Late Cretaceous could have been more than twice that at present.

The idea of Kerrick et al. (2003) can be tested using published data on the abundance of not just andesites but total terrestrial and submarine volcanics extruded over Phanerozoic time (Ronov, 1993). Total volcanics should give a better idea of total global degassing. To correct for loss by erosion over time, the data can be fitted to an exponential decay curve, just as was done for sandstones and shales in chapter 2 (equation 2.2):

$$\Delta V/\Delta t = (\Delta V/\Delta t)_o \exp(-k\tau) \tag{4.2}$$

where:

τ = mean age of a given volume of rocks extruded within the time span Δt
ΔV = volume of volcanic rocks within age span Δt
$(\Delta V/\Delta t)_o$ = rate of volcanic rock extrusion at present
k = weathering and erosion decay coefficient for volcanics (assumed constant).

Values of $\Delta V/\Delta t$ are determined from the data of Ronov (1993) for 27
age spans ranging from the Miocene to the lower Cambrian. A plot fitted to the Ronov data is shown in figure 4.1. To avoid overly biasing
the fitted curve by including excessive total rock abundance accompanying the Plio-Pleistocene glaciation, "present" is assumed to represent
the mid-Miocene (15 Ma). Also, use of a constant value for k assumes
that the probability of erosive loss does not change with time. Deviations of each time span data point above and below the exponential curve
can then be interpreted as original increases or decreases in global volcanism relative to that at present (Wold and Hay, 1990). In other words:

$$f_V(\tau) = [\Delta V/\Delta t_{ron}/\Delta V/\Delta t_{exp}] \qquad (4.3)$$

where the subscript v refers to volcanics, ron refers to the data of Ronov,
and exp to the value of $\Delta V/\Delta t$ calculated from equation (4.2) for the same
time. A plot of $f_V(t)$ derived in this manner is shown in figure 4.2. Values of $f_{SR}(t)$ used in the GEOCARB II and III models, based on the data
of Gaffin (1987) and that of Engebretson et al. (1992) (for the past
150 Ma), are also included in figure 4.2. The rapid variations in $f_V(t)$ at
some times, I believe, are due more to the inability to accurately quan-
tify ancient rock abundance than to real changes in extrusion and degassing rates. At any rate, the two approaches to paleo-degassing rate
crudely agree in the sense that they both give f values within the range
0.5–2.0.
    In the absence of a better quantitative model for global degassing over
geologic time, change in seafloor spreading rate is still probably the best

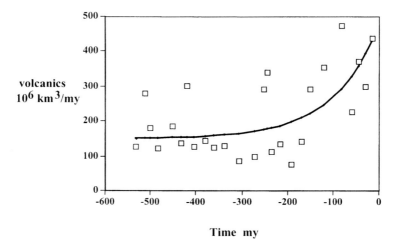

Figure 4.1. Plot of volcanic rock abundance versus time fitted by an eponential decay
(equation 4.2) representing expected loss by erosion. (Data from Ronov, 1993.)

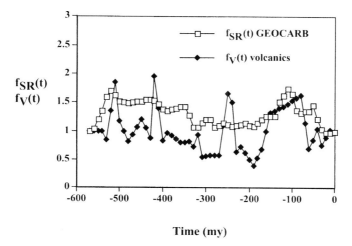

Time (my)

Figure 4.2. Plot of volcanic degassing parameters versus time. The parameter $f_{SR}(t)$ (Berner, 1994; Berner and Kothavala, 2001) is seafloor spreading rate at a past time divided by the spreading rate at present. The parameter $f_V(t)$ is the rate of volcanic eruption divided by that for the Miocene (as a representation of the "present"). Values of $f_V(t)$ are calculated from volcanic rock abundance data (Ronov, 1993) using equation (4.3).

first approximation of change in global degassing. If so, it is imperative to know how this rate has changed over geologic time. Estimates for spreading rate over the past 150 Ma have varied from essentially constant (Parsons, 1982; Heller et al., 1996; Rowley, 2002) to increases of a factor of 1.3–2 (Hays and Pitman, 1973; Pitman, 1978; Komincz, 1984; Engebretson et al., 1992; Gaina et al., 2003). Most carbon cycle models have used the Engebretson et al. formulation (which is based on subduction rates). This formulation has been challenged by Rowley (2002). However, the study by Gaina et al. (2003), which accounts for the areas of seafloor already subducted (not considered by Rowley), shows a definite overall reduction (about 50%) in spreading rate since the Jurassic. Considering the extreme amount of work, including use of marine gravity data, magnetic anomalies, bathymetry, seismic data, and geologic subduction data, that has gone into the Gaina et al. formulation, I believe it is the best one available at the present time.

The oldest seafloor is only 180 Ma. This means that there is no direct measure of spreading rate before this time. A proxy that has been used to extend calculations of global spreading rate is the history of sea level. Studies by Hays and Pitman (1973) and Kominz (1984) have demonstrated a good correlation between sea level and spreading rate for the past approximately 100 million years. This correlation comes about because an increase in the total volume of the mid-ocean ridges, which displaces seawater upward, accompanies the faster addition of new oceanic crust (i.e., faster spreading rate). If this correlation applies to earlier times, then spreading rate can be deduced from estimates of paleo-sea level. This has been done by Gaffin (1987), and his spreading rate estimates, combined with the first-order sea-level data of Vail et al. (1977), are used in the GEOCARB models (Berner, 1991, 1994; Berner and Kothavala, 2001) to calculate spreading rate before 150 Ma (figure 4.2).

The explanation of changes in sea level in terms of seafloor spreading rate has been challenged by Heller et al. (1996). They maintain that there has been little change in spreading rate, at least over the past 150 million years, and that the high sea levels of the Jurassic and Cretaceous are a result of the rifting and breakup of the supercontinent Pangaea. The filling of proto-oceanic rifts with sediment is equivalent to the enlargement of the area of the continents at the expense of the oceans (Heller and Angevine, 1985), and decreasing ocean area, for a given volume of water, must result in a rise of sea level. An additional factor is that creation of new seafloor, as a result of continental breakup, results in a decreasing mean age of oceanic crust (Worsley et al., 1984; Heller and Angevine, 1985). A younger seafloor is a hotter seafloor; a hotter seafloor is a higher seafloor (Parsons and Sclater, 1977), and a higher seafloor means higher sea level.

Metamorphic and Diagenetic CO_2 Degassing

Often the discussion of global CO_2 degassing has concentrated only on volcanic processes. This neglects additional processes that may be just as important globally. For example, the range shown in table 4.1 for estimated total mantle degassing can be distinctly lower than that estimated for CO_2 consumption by silicate weathering ($6 \pm 3 \times 10^{18}$ mol/my). The remainder must be metamorphic or diagenetic in origin (Kerrick, 2001). The pathfinding study of Barnes et al. (1978) pointed to the importance of nonvolcanic degassing in the circum-Pacific and Mediterranean-Tethyan orogenic belt. They suggested that CO_2 discharges from hot springs, gas vents, and so on, are aided by the high seismicity in these zones. The study of Kerrick (2001) emphasizes the importance of areas of high heat flow overlying subsurface magmas which includes many of the world's geothermal systems. Morner and Etiope (2002) have summarized data for measured fluxes of metamorphic and diagenetic CO_2 and CH_4 degassing from terrestrial environments and estimate large global fluxes from the land. However, their total land flux (14×10^{18} mol/my) is probably too high for application over extended geologic time because it exceeds the maximum total global degassing based on silicate weathering ($6 \pm 3 \times 10^{18}$ mol/my).

Kerrick and Caldeira (1998) and Kerrick (2001) have examined in detail the problem of metamorphic degassing. They conclude that (1) much metamorphic degassing occurs as a result of continent–continent and arc–

continent collisions; (2) extensional regimes, such as rift zones, with high heat flow are also excellent environments for degassing; (3) major faults may provide conduits for significant release to the atmosphere of deeply buried CO_2; (4) synmetamorphic intrusions may cause significant regional fluid flow, which aids decarbonation and transport of CO_2 to the surface; and (5) high CO_2 levels during the Eocene were more likely caused by degassing from the American Cordilleran belt than from the Himalaya-Karakoram belt as previously believed. In an earlier study based on measurements of degassing at the Salton Trough (California) and Taupo Volcanic Zone (New Zealand), Kerrick et al. (1995) concluded that present global metamorphic degassing could be as much as 1×10^{18} mol/my, which is the value shown in table 4.1.

During subduction at active plate boundaries, the extent of metamorphic decomposition and degassing, based on phase equilibrium arguments, depends on the thermal gradient and the ability of water to reach great depths (Kerrick and Connolly, 2001). Kerrick and Connolly conclude that little metamorphic degassing occurs at these plate boundaries, but there is not general agreement on this subject. The isotopic analysis of volcanic gases ($^{13}C/^{12}C$, ^3He) sampled at arc volcanoes indicates a considerable flux of CO_2 from subducted carbonates of 2.3×10^{18} mol/my (Sano and Williams, 1996), whereas high-pressure experiments and modeling of metamorphism suggest that most of the down-going $CaCO_3$ is subducted deeply into the mantle without metamorphic decarbonation (Molina and Poli, 2000).

Another, relatively neglected, degassing process is that accompanying both shallow and deep diagenesis. It is well known that the maturation of buried organic matter involves the loss of CO_2 (and CH_4; see below; Durand, 1980). In addition, there is the globally important process of the diagenetic conversion of smectite to illite (Hower et al., 1976). This involves the release of acid and the dissolution and removal of calcium carbonate, and it must be an additional source of low-temperature CO_2. A large variety of measurements of fluxes of CO_2 and CH_4, emanating from diagenetic processes below the land surface (including petroleum maturation), is summarized by Morner and Etiope (2002). The gases are emitted to the surface in cool seeps, springs, and so on, and the CH_4 becomes rapidly oxidized to CO_2 in the atmosphere.

Carbonate Deposition and Degassing

Any transfer of carbonate deposition from shallow-water shelves and platforms to the deep sea should have resulted in increased rates of global degassing in the geological past. This is because deep sea carbonates rest on oceanic crust that undergoes subduction and possible degassing at active plate boundaries. In chapter 3 it was shown that the rise and spread of calcareous plankton, chiefly coccoliths and forams, should have caused a shift of carbonate burial from shallow to deep water starting at about 150 Ma in the Jurassic. This shift has been demonstrated quantitatively by Wilkinson and Walker (1989) based on rock abundance data. The effect on CO_2 degassing over the past 150 million years depends on how fast the shift has occurred, whether it has been accelerating or decelerating, and whether the pelagic carbonate has been subducted and, if subducted, also decarbonated and degassed. For example, present pelagic $CaCO_3$ accumulation is much higher than the rate of $CaCO_3$ subduction because a large $CaCO_3$ fraction accumulates on the Atlantic seafloor, which is surrounded by passive continental margins (Schrag, 2002). Berner (1991) originally assumed a decelerating transfer of carbonate to the deep-sea floor based on DSDP (Deep Sea Drilling Project) paleontological data (Hay et al., 1988), but the rock abundance data of Wilkinson and Walker has caused a linerarly increasing effect to be assumed in later work (Berner, 1994; Berner and Kothavala, 2001). In the latter case the effect on degassing of the shift from shallow water to deep water deposition, resulting in increasing carbonate in subducting crust, is represented by the dimensionless parameter $f_C(t)$:

For the past 150 million years:

$$f_C(t) = f_C(150) + \{[1 - f_C(150)]/150\}(150 - t) \qquad (4.2)$$

Before 150 Ma:

$$f_C(t) = f_C(150) \qquad (4.3)$$

where $f_C(t)$ is the effect of carbonate content of subducting oceanic crust on the rate of CO_2 degassing.

The constant value at and before 150 Ma, $f_C(150)$, is assumed to represent only nonsubduction zone degassing plus subduction of only that carbonate that infills basaltic crust because of the absence of convincing data for the existence of pelagic deep sea carbonates before this time (see chapters 2 and 3).

Methane Degassing

Methane is a potent greenhouse gas, 30 times stronger per molecule than CO_2, and it is produced during both the short-term and long-term carbon cycles. In the (prehuman) short-term cycle, it is produced mainly from wetlands and animal exhalation such as from bovids and termites. (These organisms have bacteria located in their digestive systems that break down carbohydrates to methane.) In wetlands and other water-logged, organic-rich sediments, methane forms from a variety of microbial processes and chemical pathways, but the overall reaction can be simplified as

$$2CH_2O \rightarrow CO_2 + CH_4 \tag{4.4}$$

Because methane is readily oxidized by dissolved O_2 (see table 4.2), it cannot accumulate in the presence of O_2, and that is why appreciable methane is found only in water-logged sediments where all dissolved O_2 has been previously removed by microbial respiration. Also, methane forms preferentially in fresh water sediments, as opposed to marine sediments, because methanogenic microorganisms are outcompeted for organic substrate by other microbes using dissolved sulfate as an energy source (Claypool and Kaplan, 1974), and marine sediments are rich in interstitial sulfate. As a result, biogenic methane formation occurs in marine sediments only after the removal of O_2 and sulfate by other microbes (Claypool and Kaplan, 1974; Froelich et al., 1979). The order of the pathways by which organic matter decomposition takes place follows changes in free energy (table 4.2).

In the long-term carbon cycle most methane is produced at depth in sediments by both biogenic and abiogenic organic matter decomposition. At low temperatures biogenic decomposition occurs via the same reaction as (4.4). In buried freshwater organic-rich sediments, such as peat, biogenic methane produced at depth can readily move upward through the sediment and escape to the overlying water and, if not oxidized to CO_2 by O_2 in the water, eventually to the atmosphere. By contrast, in buried marine sediments upward-diffusing methane is oxidized by reaction with interstitial dissolved sulfate (Martens and Berner, 1977; Reeburgh and Heggie, 1977; Valentine et al., 2002), which is essentially absent in the interstitial waters of freshwater sediments. Because the sulfate concentration in seawater is so high (> 100× the average for fresh waters), methane flux out of marine sediments is limited. The free energy yield for CH_4 reacting with sulfate under sedimentary conditions shows that this reaction is thermodynamically favored (table 4.2).

Table 4.2. Some major microbially mediated organic matter decomposition reactions with standard-state free-energy change, ΔG°, and free-energy change for typical activities of gases and dissolved species in organic-rich marine sediments during early diagenesis, ΔG^*.

Reaction	ΔG° (kJ/mol)	ΔG^*
$CH_2O + O_2 \rightarrow CO_2 + H_2O$	−475	−491
$2CH_2O + SO_4^{-2} \rightarrow H_2S + 2HCO_3^-$	−77	−110
$2CH_2O \rightarrow CO_2 + CH_4$	−58	−44
$CH_4 + 2O_2 \rightarrow CO_2 + 2H_2O$	−581	−594
$CH_4 + SO_4^{-2} + CO_2 \rightarrow H_2S + 2HCO_3^-$	−14	−25

Activity for CH_4 assumed to be saturation with gas at 25°C; CH_2O represented by sucrose. Data partly from Berner (1971).

By comparison with biogenic methane, abiogenic methane forms at much greater depths due to the thermal decomposition of the organic matter surviving bacterial decomposition. This remaining organic matter is normally referred to as kerogen. As kerogen matures, it loses CH_4, CO_2, and H_2O (Durand, 1980) and trends toward pure carbon (graphite) at the highest metamorphic temperatures (several hundreds of degrees Celsius). Thermal maturation of kerogen gives rise to the formation of coal, oil, and natural gas (CH_4), depending on the temperature, time of burial, and the nature of the starting organic material. If the natural gas can move into fractures leading to the earth's surface, it may escape into the atmosphere.

Another minor source of abiogenic methane is at mid-ocean ridge (MOR) hydrothermal vents (e.g., Welhan and Craig, 1979). The methane apparently results from the reaction of CO_2 with H_2 formed by the reduction of water accompanying iron mineral oxidation. Although this reaction has been found to be associated with the overall process of the serpentinization of ultramafic rocks, additional methane can form from the hydrothermal alteration of MOR basalts (Welhan, 1988).

Even though methane is a potent greenhouse gas, it does not attain levels in the present atmosphere sufficiently high enough to overshadow warming due to CO_2. This is also likely for most of the Phanerozoic. If the relative abundance of coal swamps over time can be assumed to represent the relative rate of input of methane to the atmosphere (Berner and Mackenzie, unpublished ms), and levels of atmospheric O_2 have not dropped below half the present level (Chaloner, 1989), then past levels of atmospheric CH_4 could not have been high enough to dominate greenhouse warming. This is because CH_4 is rapidly oxidized in the atmosphere by O_2 to CO_2 (mean residence time of about 10 years). Thus, the idea that over many millions of years high levels of atmospheric CH_4 brought about sustained global warming in the past is difficult to maintain.

Methane can arise from yet another source. This is via the breakdown of methane hydrates. At sufficiently low temperatures and high pressures, methane will react with water to form crystalline hydrate phases known as clathrates. Because they are thermodynamically unstable under earth surface conditions or at great depths when termperatures become too high, the clathrates are found only at intermediate depths in marine and nonmarine sediments. The stability of the hydrates in seawater is shown in figure 4.3. For clathrates occurring in the uppermost portions of sediments this means water depths of 200 m at 0°C and 1000 m at 12°C. However, most clathrates are found at considerable depth in the sediment column. The depth range depends on gas concentration and the available pore space within certain stability limits. Conditions neccessary to stabilize hydrates in marine sediments are pressures of 3–5 MPa at 3°C and 8–12 MPa at 11°C (Østergaard et al.,

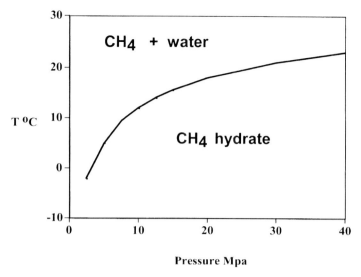

Figure 4.3. Pressure–temperature stability of methane hydrate in seawater. (Data from Peltzer and Brewer, 2000.)

2002), with the range of pressures representing in each case the effects of fine particles in stabilizing the hydrates. Potentially, gas hydrates can occur between the seafloor and a locus of sub-bottom depths where geothermal gradients intersect gas–gas hydrate-pore water equilibrium curves (Dickens, 2001). Important controls are gas composition, water activity, bottom water temperature, geothermal gradient, and water depth. This means that the clathrates are found either near the sediment surface in cool and moderately deep-water marine sediments or within a limited depth range in the sediment column. Occurrences in marine sediments are worldwide, and they are also found buried in thick permafrost soils in polar regions (Kvenvolden, 1993, 2002).

Because methane reacts with both dissolved O_2 and sulfate, both must be absent for CH_4 to build up to saturation with clathrates in marine sediments. This is not normally a problem in organic-rich sediments. However, upward moving methane gas, released from hydrates, must travel through an overlying zone where it can be removed by bacterial sulfate reduction (Reeburgh, 1983; Valentine et al., 2001). Also, the methane must pass rapidly through overlying oxygenated waters to avoid further removal via oxidation before it can be transferred to the atmosphere (Dickens, 2000; Valentine et al., 2001). The best places for clathrates to form so that their potential decomposition can give rise to a flux of CH_4 out of the sediment is at shallow sediment depths in unusually organic-rich sediments, where sulfate reduction is very intense and all interstitial sulfate is removed over a short depth range. However, the sediments still have to be sufficiently cold and under sufficiently deep water for clathrates to have formed in the first place.

What triggers methane hydrates to break down to methane gas which escapes to the atmosphere? Some possible causes are submarine landslides, sea level changes, and changes in ocean circulation (Bice and Marotzke, 2002). The sudden release of large quantities of methane from clathrates has been used to explain several short-lived events during Phanerozoic history. This includes the late Paleocene/Eocene thermal maximum (e.g., Dickens et al., 1997), Mesozoic oceanic anoxic events (e.g., Hesselbo et al., 1990; Beerling et al., 2002; Bralower et al., 2002), an early Cretaceous episode of atmospheric CO_2 increase (Jahren et al., 2001), the Permo-Triassic extinction (Krull and Retallack, 2000; McLeod et al., 2000), and the Triassic-Jurassic extinction (Beerling and Berner, 2002). The evidence in the sedimentary record for sudden methane release is a rapid decrease in the carbon isotopic composition of sedimentary $CaCO_3$ and organic matter. This comes about because methane is unusually light; in other words, it is highly depleted in ^{13}C (average $\delta^{13}C$ values are $-60‰$ to $-80‰$). After a large mass of isotopically light methane is released to the atmosphere, it is oxidized rapidly to CO_2 and then cycled by weathering and sedimentation to form unusually light carbonates and organic matter that are deposited in sediments. No other reasonably sized source of light carbon has yet been suggested to explain large, short-term negative isotope anomalies (see chapter 5, figure 5.15). Besides negative anomalies, some positive anomalies may be due to methane hydrates. It has been suggested that some positive excursions in the geologic record may have been caused by the formation and storage of methane hydrates rather than the burial of excess organic matter (Dickens, 2003).

Summary

It should be apparent from the discussion in this chapter that there are many unanswered questions regarding the rate of CO_2 degassing over geologic time, including the Phanerozoic. First, there is a need for better data on present degassing rates, especially on the relative contributions of metamorphism and diagenesis to global degassing. Second, there is the problem of how degassing rate has changed over geologic time. If seafloor creation (spreading) rate is an important consideration, then there is a need for better data on this subject. If spreading is not a good degassing indicator, then other paleo-degassing methods are needed. One possibility is the abundance of volcanic rocks, a topic addressed in this chapter by means of a new calculation of relative degassing based on Russian data on volcanic rock abundance. Even if done correctly, however, volcanic rock abundance only accounts for volcanic degassing, and not that due to metamorphism and diagenesis. Third, degassing accompanying regional metamorphism in compressional mountain belts needs further study. Fourth, hot-spot and md-plate volcanism needs better integration into carbon cycle models. Finally, changes over time in the carbonate content of down-going oceanic slabs during subduction need to be better quantified. This leads to such questions as, were there abundant pelagic carbonates during the Paleozoic? Of all the major aspects of the Phanerozoic carbon cycle, the input process of global CO_2 degassing is the least well understood.

It is unlikely that methane has served as a principal greenhouse gas over long periods during the Phanerozoic. However, on the scale of less than a million years, the release of methane from the breakdown of methane hydrates may have caused dramatic excursions in temperature driven by excessive CO_2 formed by the rapid atmospheric oxidation of the released methane. Large short-term negative carbon isotopic excursions are best explained by this process.

5

Atmospheric Carbon Dioxide over Phanerozoic Time

In this chapter the methods and results of modeling the long-term carbon cycle are presented in terms of predictions of past levels of atmospheric CO_2. The modeling results are then compared with independent determinations of paleo-CO_2 by means of a variety of different methods. Results indicate that there is reasonable agreement between methods as to the general trend of CO_2 over Phanerozoic time.

Long-Term Model Calculations

Values of fluxes in the long-term carbon cycle can be calculated from the fundamental equations for total carbon and ^{13}C mass balance that are stated in the introduction and are repeated here:

$$dM_c/dt = F_{wc} + F_{wg} + F_{mc} + F_{mg} - F_{bc} - F_{bg} \qquad (1.10)$$

$$d(\delta_c M_c)/dt = \delta_{wc}F_{wc} + \delta_{wg}F_{wg} + \delta_{mc}F_{mc} \qquad (1.11)$$
$$+ \delta_{mg}F_{mg} - \delta_{bc}F_{bc} - \delta_{bg}F_{bg}$$

where

   M_c = mass of carbon in the surficial system consisting of the atmosphere, oceans, biosphere, and soils

F_{wc} = flux from weathering of Ca and Mg carbonates
F_{wg} = flux from weathering of sedimentary organic matter
F_{mc} = degassing flux for carbonates from volcanism, metamorphism, and diagenesis
F_{mg} = degassing flux for organic matter from volcanism, metamorphism, and diagenesis
F_{bc} = burial flux of carbonates in sediments
F_{bg} = burial flux of organic matter in sediments
$\delta = [(^{13}C/^{12}C)/(^{13}C/^{12}C)stnd - 1]1000.$

Variants of equations (1.10) and (1.11) have been treated in terms of non–steady-state modeling (e.g., Berner et al., 1983; Wallmann, 2001; Hansen and Wallmann, 2003; Mackenzie et al., 2003; Bergman et al., 2003), where the evolution of both oceanic and atmospheric composition, including Ca, Mg, and other elements in seawater, is tracked over time. However, since the purpose of this book is to discuss the carbon cycle with respect to CO_2 and O_2, and so as not to overburden the reader with too many mathematical expressions, I discuss only those aspects of the non–steady-state models that directly impact carbon. These are combined with results from steady-state strictly carbon-cycle modeling (Garrels and Lerman, 1984; Berner, 1991, 1994; Kump and Arthur, 1997; Francois and Godderis, 1998; Tajika, 1998; Berner and Kothavala, 2001; Kashiwagi and Shikazono, 2002). For steady state, we have (from chapter 1):

$$dM_c/dt = 0; \quad d(\delta_c M_c)/dt = 0 \tag{1.12}$$

and

$$F_{wsi} = F_{bc} - F_{wc} \tag{1.13}$$

where F_{wsi} represents the weathering of Ca and Mg silicates with the transfer of carbon from the atmosphere to Ca and Mg carbonates (reaction 1.4).

In all the long-term carbon cycle models, values of carbon fluxes and the concentration of atmospheric CO_2 are obtained by combining variants of equations (1.10) and (1.11), with additional expressions for fluxes that include nondimensional weathering and degassing parameters. An example of this approach, taken from steady-state GEOCARB modeling (Berner, 1991, 1994; Berner and Kothavala, 2001), is:

$$F_{wsi}(t) = F_{bc}(t) - F_{wc}(t) \tag{5.2}$$
$$= f_{Bt}(CO_2)f_{Bb}(CO_2)\ f_R(t)\ f_E(t)f_{AD}(t)^{0.65}\ F_{wsi}(0)$$

$$F_{wc}(t) = f_{Bc}(CO_2)f_{Bb}(CO_2)\ f_E(t)f_{AD}(t)f_{LA}(t)\ k_{wc}\ C \tag{5.3}$$

$$F_{wg}(t) = f_R(t) \, f_D(t) \, k_{wg} G \qquad (5.4)$$

$$F_{mc}(t) = f_{SR}(t) \, f_C(t) \, F_{mc}(0) \qquad (5.5)$$

$$F_{mg}(t) = f_{SR}(t) \, F_{mg}(0) \qquad (5.6)$$

where

C, G = global mass of sedimentary carbonate carbon and organic carbon

k_{wc}, k_{wg} = rate constants = $F_{wc}(0)/C$, $F_{wg}(0)/G$ for the present earth

$f_{Bc}(CO_2) = f_{Bt}(CO_2)$ with different Z values for carbonate dissolution

$f_{AD}(t) = f_D(t) f_A(t)$

$f_{LA}(t) = f_L(t) f_A(t)$.

$F_{wsi}(0)$, $F_{mc}(0)$, and $F_{mg}(0)$ refer to the present earth, and the other dimensionless f parameters are defined and discussed in earlier chapters. For convenience, a summary of definitions of all terms is shown in table 5.1. In GEOCARB modeling, the concentration of atmospheric CO_2 at each time is calculated by solving equation (5.2) for $f_B(CO_2) = [f_{Bt}(CO_2) f_{Bb}(CO_2)]$ (or for the prevascular plant land surface, $f_B(CO_2) = [f_{Bt}(CO_2) f_{Bnb}(CO_2)]$) and then by inverting the resulting complex expression to solve for RCO_3, the ratio of the mass of atmospheric CO_2 at some past time to that for the preindustrial present (280 ppmv). A plot of RCO_2 versus $f_{Bt}(CO_2)$ from GEOCARB III (Berner and Kothavala, 2001) is shown in figure 5.1. The plot shows sudden variations, not seen in the smooth plot of RCO_2 versus the greenhouse parameter $f_{Bg}(CO_2)$ (see figure 2.5), because of variations over time of GEOG, the effect of changing continent size and distribution on land temperature.

To be able to solve the above equations, it is necessary to track C and G over time. If total crustal carbon is assumed to be constant (in other words, the loss of carbon to the mantle equals its gain from the mantle), then C + G = constant, and

$$dC/dt = F_{bc} - F_{wc} - F_{mc} \qquad (5.7)$$

$$dG/dt = F_{bg} - F_{wg} - F_{mg} \qquad (5.8)$$

Analogous expressions for ^{13}C are

$$d(\delta_c C)/dt = \delta_{bc} F_{bc} - \delta_{wc} F_{wc} - \delta_{mc} F_{mc} \qquad (5.9)$$

$$d(\delta_g G)/dt = \delta_{bg} F_{bg} - \delta_{wg} F_{wg} - \delta_{mg} F_{mg}. \qquad (5.10)$$

For simplification one can assume that $\delta_c = \delta_{wc} = \delta_{mc}$ and $\delta_g = \delta_{wg} = \delta_{mg}$. An additional simplification is the assumption that δ_{bc} represents the

Table 5.1. Definitions of terms used in GEOCARB long-term carbon cycle modeling.

Fluxes, masses and isotopic composition

F_{wc} = carbon flux from weathering of Ca and Mg carbonates
F_{wg} = carbon flux from weathering of sedimentary organic matter
F_{mc} = degassing flux from volcanism, metamorphism, and diagenesis of carbonates
F_{mg} = degassing flux from volcanism, metamorphism, and diagenesis of organic matter
F_{bc} = burial flux of carbonate-C in sediments
F_{bg} = burial flux or organic-C in sediments
C = mass of crustal carbonate carbon
G = mass of crustal organic carbon
k_{wc} = rate constant expressing mass dependence for carbonate weathering
k_{wg} = rate constant expressing mass dependence for organic matter weathering
δ_{wc} = $\delta^{13}C$ of flux from weathering of Ca and Mg carbonates
δ_{wg} = $\delta^{13}C$ of flux from weathering of sedimentary organic matter
δ_{mc} = $\delta^{13}C$ of degassing flux from carbonate decomposition due to volcanism, metamorphism, and diagenesis
δ_{mg} = $\delta^{13}C$ of degassing flux from organic decomposition due to volcanism, metamorphism, and diagenesis
δ_{bc} = $\delta^{13}C$ of burial flux of carbonates in sediments (assumed same as $\delta^{13}C$ of ocean)
δ_{bg} = $\delta^{13}C$ of burial flux of organic matter in sediments

---
Dimensionless parameters

RCO_2 = mass of CO_2 in atmosphere (t)/mass of CO_2 in atmosphere (0)
$f_R(t)$ = effect of relief on weathering rate (t)/effect of relief on weathering rate (0)
$f_E(t)$ = effect of plants on weathering rate (t)/effect of plants on weathering rate (0)
$f_{Bb}(CO_2)$ = effect of CO_2 on plant-assisted weathering (t)/effect of CO_2 on plant-assisted weathering (0) for silicates and carbonates
$f_{Bnb}(CO_2)$ = direct effect of atmospheric CO_2 on weathering rate (t)/direct effect of atmospheric CO_2 on weathering rate (0) for silicates and carbonates (applied to period before rise of large land plants)
$f_{Bg}(CO_2)$ = effect of temperature on weathering rate (t)/effect of temperature on weathering rate (0) for silicates due to CO_2 greenhouse effect alone
$f_{Bt}(CO_2)$ = effect of temperature on weathering rate (t)/effect of temperature on weathering rate (0) for silicates combining effects of CO_2 greenhouse, solar radiation, and paleogeography on temperature
$f_{Bc}(CO_2)$ = effect of temperature on weathering rate (t)/effect of temperature on weathering rate (0) for carbonates combining effects of CO_2 greenhouse, solar radiation, and paleogeography on temperature
$f_D(t)$ = runoff (t) / runoff (0) due to changes in paleogeography
$f_A(t)$ = land area (t)/land area (0).
$f_L(t)$ = fraction of total land area covered by carbonates/the same fraction at present
$f_{SR}(t)$ = seafloor area creation rate (t)/seafloor area creation rate (0)
$f_C(t)$ = effect of carbonate content of subducting oceanic crust on the rate of CO_2 degassing (t)/the same effect at present

---

From Berner (1991, 1994); Berner and Kothavala (2001).

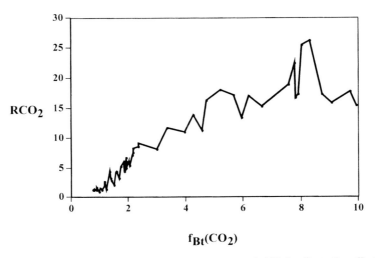

$$f_{Bt}(CO_2)$$

Figure 5.1. Plot of $f_{Bt}(CO_2)$ versus RCO_2. The parameter $f_{Bt}(CO_2)$ reflects the effect on the rate of silicate weathering and atmospheric CO_2 of changes over time in temperature, due to the CO_2 greenhouse effect, changes in solar radiation, and changes in mean land temperature accompanying continental drift. RCO_2 is the ratio of the mass of atmospheric CO_2 at a past time (t) to that at present (assumed to be 280 ppm). The irregularities in the curve (compare with figure 2.5) are due primarily to variations of the GEOG term for mean land temperature.

carbon isotopic composition of the oceans (literally, it is the mean composition of carbon in buried carbonates) and that $\delta_{bg} - \delta_{bc} = \Delta^{13}C$, which can be held constant or varied over time.

Combining the above expressions constitutes GEOCARB modeling. By inserting values for present-day masses and fluxes, using the carbon isotopic composition of the oceans plus values of the dimensionless f parameters over the Phanerozoic, and by assuming starting values (at 550 Ma) for fluxes, masses, and isotopic composition, a series of steady-state equations for the past 550 million years is solved at each 1 million year time-step for all fluxes and for RCO_2. The assumed starting values are validated only if the present CO_2 level is obtained. As values for the various dimensionless parameters are inputted only every 10 million years, with linear interpolation during the time gaps, output results of the modeling are reported on the same basis. This precludes consideration of any phenomenon occurring on a time scale shorter than 10 million years.

The carbon cycle modeling approach of other studies (Kump and Arthur, 1998: Francois and Godderis, 1998; Tajika, 1998; Wallmann, 2001; Kashiwagi and Shikazono, 2002) is fundamentally similar to that for GEOCARB. However, modifications of the fluxes and dimensionless parameters (f) are made, and these modifications are worth discussing in some detail. Kump and Arthur (1997), in a model for the Cenozoic carbon cycle, use the parameter $f_{wr}(t)$ to represent all other factors affecting weathering other than $f_L(t)$ and $f_A(t)$. In essence they are combining the effects of both climate, $f_{Bt}(CO_2)$, paleogeographically-affected runoff, $f_D(t)$, and mountain uplift/erosion, $f_R(t)$, into their $f_{wr}(t)$ parameter. Francois and Godderis (1998), in their model for the Cenozoic, join carbon cycle modeling with modeling of Sr isotopes and greatly simplify weathering such that carbonate weathering, F_{wc}, and organic matter weathering, F_{wg}, are assumed to be directly proportional to silicate weathering, F_{wsi}.

Tajika (1998), in a model for the past 150 Ma, assumes minor loss of carbon to the mantle via subduction and divides the global CO_2 degassing flux $(F_{mc} + F_{mg})$ into separate fluxes at mid-ocean ridges, that associated with intraplate (hot-spot) volcanism, and that accompanying metamorphism and volcanism at subduction zones. Mid-ocean ridge and subduction zone degassing is assumed proportional to seafloor spreading rate, $f_{SR}(t)$, whereas for hot-spot degassing over time he introduces a new parameter, $f_H(t)$, which varies differently from spreading rate and relies on the data of Larson (1991). Tajika also separates the global weathering flux into that for the Himalayas (including the uplift parameter $f_R(t)$ based on Sr isotopes) and that for the rest of the world.

In his carbon cycle model for the past 150 Ma, Wallmann (2001), like Tajika (1998), separates degassing into that at ridges, subduction zones, and hot spots with degassing at ridges and hot spots guided by spreading rate and that at hot spots guided by the data of Larson (1991). Unlike GEOCARB, Wallmann's model emphasizes the submarine weathering of basalt to $CaCO_3$. Other differences are that he considers loss of carbon to the mantle, calculates the carbon isotopic composition of the oceans over time (rather than using equation 1.11), and separates Ca and Mg silicate weathering into young volcanics at convergent margins and silicates from the rest of the world. Kashiwagi and Shikazono (2002), in a model for the Cenozoic, distinguish subduction zone degassing at island arcs from that at back-arc basins, and they, like Wallmann, separate weathering of silicates for the Himalayas from that for the rest of the world.

Perturbation Modeling

Sudden and rapid perturbations of the long-term carbon cycle can bring about changes requiring the direct use of non–steady-state modeling. Some examples are the effect on the level of atmospheric CO_2 and oceanic $\delta^{13}C$ of instant decarbonation of limestones accompanying extraterrestrial bolide impacts (Caldeira and Rampino, 1990; Kump, 1991; Beerling et al., 2002), degassing of CO_2 accompanying massive volcanic eruptions (Dessert et al., 2001), and degassing of CH_4 from the decomposition of methane hydrates followed immediately by oxidation of CH_4 to CO_2 (e.g., Dickens, 1997; Beerling and Berner, 2002; Berner, 2002). Sudden perturbation of the steady-state carbon cycle normally takes about 500,000 years to return to a new steady state (Sundquist, 1991). Over such a "short" time, more slowly varying processes can be ignored, including changes in the position, size, relief, and exposed lithology of land areas. This is equivalent to holding $f_D(t)$, $f_A(t)$, $f_R(t)$, and $f_L(t)$ constant. Degassing parameters, such as $f_G(t)$, $f_C(t)$, which also change only very slowly, are ignored and replaced by jumps in $F_{mc}(t)$ and $F_{mg}(t)$ that are either instantaneous (bolide impacts) or occur by square wave or Gaussian input functions over periods of thousands to tens of thousands of years (methane hydrate degassing) to a few hundred thousand years (volcanic degassing). Also, the surficial reservoir is divided into its components, and separate carbon mass-balance expressions are used for the oceans, atmosphere, and biosphere. In other words, short-term carbon cycling is mixed with long-term cycling. Furthermore, where carbonate precipitation is involved, mass balance expressions for Ca, Mg, and alkalinity are included.

Model Results

The standard (best estimate) curve of atmospheric CO_2 for the past 150 million years from GEOCARB III (Berner and Kothavala, 2001) is compared in figure 5.2 with the results of two other recently published carbon cycle modeling studies (Tajika, 1998; Wallmann, 2001). (In the carbon cycle study of Kump and Arthur [1997], a CO_2 curve was calculated but believed by them to be incorrect, and in that of Francois and Godderis [1998], no CO_2 curve was presented.) In the overall trend of decreasing CO_2 with time, there is general agreement between the results of the three studies, but disagreements arise as to absolute RCO_2 values and shorter term variations. The Mesozoic values for GEOCARB III are higher than the results of the other two studies mainly because values of $f_E(t)$ for Mesozoic gymnosperms versus angiosperms were used in GEOCARB that best fitted the independent RCO_2 data of Ekart et al. (1999). The non-fitted results of the earlier study, GEOCARB II (Berner, 1994), better agree with those of Tajika (1998) and Wallmann (2001). Together the results of the three studies are in accord with the conclusion that overall global cooling over the past 150 million years can be explained, at least partly, by a diminished atmospheric greenhouse effect due to a decline in atmospheric CO_2.

The best estimate curve of CO_2 versus time for the entire Phanerozoic from GEOCARB III (Berner and Kothavala, 2001) is shown in figure 5.3. The most dramatic feature of the curve is the large drop in CO_2 occurring in the mid-Paleozoic (400–300 Ma). This drop is due mainly to a combination of changes brought about by the rise of large vascular

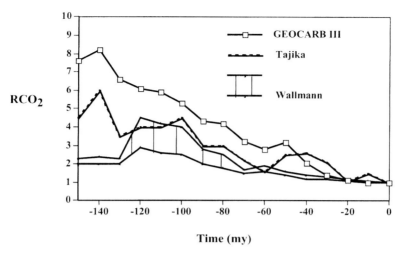

Figure 5.2. Plots of RCO_2 versus time for the theoretical modeling results of Tajika (1998), Wallmann (2001), and Berner and Kothavala (2001).

land plants (see chapter 2). The plants both accelerated weathering and provided biologically resistant organic remains for burial in sediments, causing a drop in CO_2. This major effect of plants is also shown by the modeling results of Bergman et al. (2003) and Mackenzie et al. (2003). The results of Bergman et al. are compared to those of GEOCARB III in figure 5.3. Both curves show a large drop in CO_2 during the Paleozoic, but the Bergman curve shows less variation in CO_2 from 350 to 100 Ma, with no high values during the Mesozoic as calculated by the GEOCARB model.

Carbon cycle models can give some idea of how the composition of the atmosphere and oceans may have varied over geologic time, but their chief utility is in determining sensitivity to changes in values of inputted parameters. One way of showing sensitivity is to hold constant, at the present value of 1, each dimensionless f parameter that normally varies over time. Another approach is to vary critical constants in the climate and biological feedback expressions. This procedure is illustrated in figures 5.4–5.12 using results from GEOCARB III (Berner and Kothavala, 2001).

Figure 5.4 shows the effects of holding $f_E(t) = 1$. This is equivalent to assuming that the present effect of plants on silicate and carbonate weathering rate applies to the entire Phanerozoic. An enormous effect is found for the early Paleozoic, and this shows how important the evolution of land plants has been in affecting the evolution of atmospheric CO_2 (Berner, 1998). The values of $f_E(t)$ for Ca and Mg silicate weathering before the rise of large vascular land plants can be estimated from studies of modern weathering, and the range (table 2.1) is about one-third

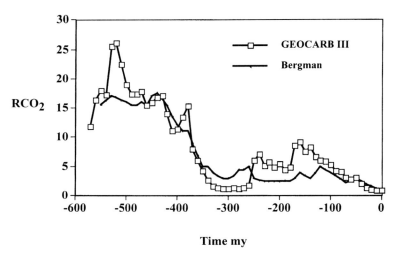

Figure 5.3. Plot of RCO_2 versus time for the standard or best estimate GEOCARB III model of Berner and Kothavala (2001) compared to the results from the model of Bergman et al. (2003).

to one-eighth. The best estimate used for the standard situation, shown in figure 5.4, is one-fourth. The high sensitivity to values of $f_E(t)$, and how they change with time, shows how important it is to obtain more data on the role of plants in weathering.

Changing paleogeographic parameters also has a major influence on the evolution of atmospheric CO_2. The effect of holding $f_R(t)$, the relief or mountain uplift factor, equal to 1 is shown in figure 5.5. The result is a dramatic drop in CO_2 during the Mesozoic and Cenozoic (past 250 million years). The effect of holding the parameters $f_D(T)$ and $f_A(t)$, which manifest the effect of paleogeography on river runoff and discharge, equal to 1, is shown in figure 5.6. This results in a much smaller effect than holding $f_R(t)$ equal to 1. The effect of letting GEOG = 0 in the expression for $f_{Bt}(CO_2)$, which is equivalent to saying that continental drift had no effect on land temperature, is shown in figure 5.7. This shows a large effect during the early Paleozoic and a large relative effect in the early Mesozoic (250–180 Ma). These results illustrate that, based on the available paleogeographic data used in the GEOCARB modeling, mountain building and changes in land temperature accompanying continental drift over Phanerozoic time had a greater effect on CO_2 than changes in land area and global runoff.

The effects of changes in values for critical constants in the weathering feedback parameters $f_{Bt}(CO_2)$ and $f_{Bb}(CO_2)$ are presented in figures 5.8–5.10. Figure 5.8 presents the effect of changing the value of Γ, the coefficient relating temperature to CO_2 via the atmospheric greenhouse effect, from the values 3.3–4.0, used in GEOCARB III, to 6.0. This

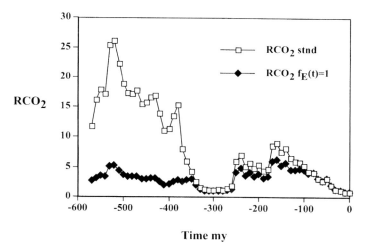

Figure 5.4. Comparison sensitivity plot of RCO_2 versus time for the situation of $f_E(t) = 1$ for all time compared to the standard GEOCARB III curve (stnd). This shows the effect on CO_2 of ignoring plant evolution and holding for all time the influence of plants on weathering to be the same as at present.

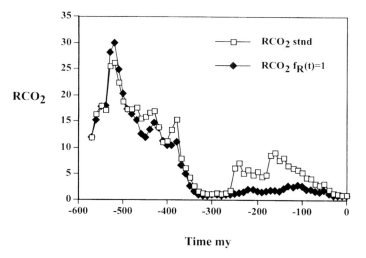

Figure 5.5. Comparison sensitivity plot of RCO_2 versus time for the situation of $f_R(t) = 1$ for all time compared to the standard GEOCARB III curve. This shows the effect on CO_2 and on weathering of holding the relief of the land, as it affects erosion, to be the same as at present.

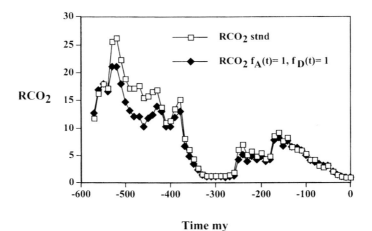

Figure 5.6. Comparison sensitivity plot of RCO_2 versus time for the situation of $f_A(t) = 1$ and $f_D(t) = 1$ so that $f_{AD}(t) = 1$ for all time compared to the standard GEOCARB III curve. This shows the effect on CO_2 and on weathering of holding both land area and river runoff to be the same as at present.

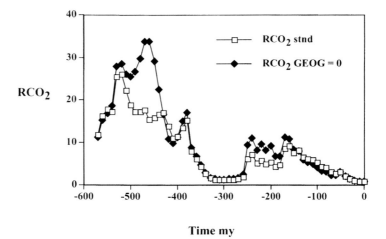

Figure 5.7. Comparison sensitivity plot of RCO_2 versus time for the situation of GEOG = 0 for all time compared to the standard GEOCARB III curve. This shows the effect on CO_2 and on weathering of holding land temperature to be the same as at present.

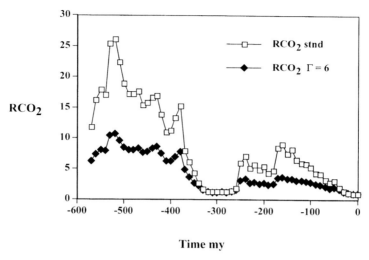

Figure 5.8. Comparison sensitivity plot of RCO_2 versus time for the situation of $\Gamma = 6$ for all time compared to the standard GEOCARB III curve. This shows the effect on CO_2 and on weathering of changing the coefficient, Γ, that relates temperature to CO_2 via the atmospheric greenhouse effect. The value of $\Gamma = 6$, higher than those values used in GEOCARB III modeling, illustrates the situation of greater sensitivity of temperature to CO_2 level and stronger negative feedback.

is equivalent to changing the temperature rise, due to a doubling of CO_2, from 2.3–2.8 to 4.1°C. This difference falls within the range of values predicted by GCM models for future global warming (IPCC, 2001). The effect of changing Γ is large for all times because it is a key component in the calculation of RCO_2. A more sensitive climate to CO_2 change means a lesser variation of CO_2 levels over time. In other words, for a larger value of Γ, there is stronger feedback and greater damping of oscillations. Varying Z, which expresses the effect of temperature on mineral dissolution rate (equation 2.11), is shown in figure 5.9. Again, there are major changes in CO_2 for all times, and Z is another important feedback damping factor. Finally, the effect of varying the exponent n in the biological weathering feedback factor f_{Bb} (CO_2) (equation 2.6) is shown in figure 5.10. Values of n = 0 and n = 1 represent the situations where no plants and all plants, respectively, increase weathering in response to increases in atmospheric CO_2. Changes in the value of n have a major effect on CO_2 in the Mesozoic and Cenozoic, but no effect during much of the Paleozoic when large plants were not present.

A special climate feedback coefficient is Ws, which reflects the effect on temperature and CO_2 of the evolution of the sun. Solar radiation has been increasing at an essentially constant rate since the early Precambrian, when it was about 30% lower than at present (Caldeira and

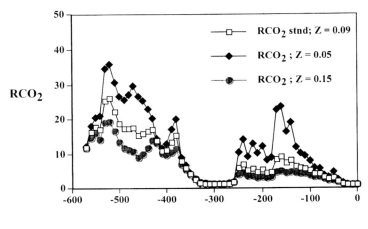

Figure 5.9. Comparison sensitivity plot of RCO_2 versus time compared to the GEOCARB III standard curve for the situation of varying Z, the coefficient that relates the rate of mineral dissolution during weathering to changes in temperature. This shows the effect on CO_2 and on weathering of varying the activation energy for dissolution (see table 2.2).

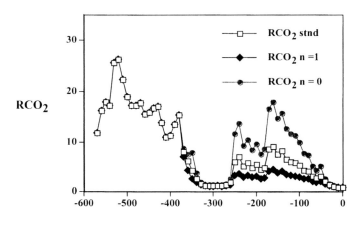

Figure 5.10. Comparison sensitivity plot of RCO_2 versus time compared to the standard GEOCARB III curve for the situation of varying n, the exponent that relates the rate of plant-assisted weathering to CO_2 fertilization of plant growth. If n = 0, no plants globally respond to CO_2 fertilization; if n = 1, all plants respond to CO_2 fertilization. The standard curve is for n = 0.4, indicating that about 35% of plants globally respond to CO_2 fertilization.

Kasting, 1992). The effect of an increase in solar radiation of about 6% over the Phanerozoic can be seen in figure 5.11. When the value of Ws in equation 2.29 is set equal to zero, signifying no change in solar radiation over time, the overall negative slope of the CO_2 curve flattens considerably. This means that the effect on CO_2 of a steady increase in solar radiation over the Phanerozoic is matched by an overall decrease in CO_2. Extrapolation of this inverse effect to the distant future indicates that the solar-driven drop in CO_2 would attain such low values in about 900 million years that photosynthesis would cease (Caldeira and Kasting, 1992).

The effect of changes in global degassing on atmospheric CO_2 is manifested in GEOCARB modeling by $f_{SR}(t)$ and $f_C(t)$, the parameters that reflect the effects of seafloor spreading (and subduction) rate and the carbonate content of subducting crust, respectively. In figure 5.12 these parameters are held constant and equal to 1 for the entire Phanerozoic. This is equivalent to assuming that there was no change in degassing rate over time and that past rates were the same as that at present. The alternative formulation of using the abundance of volcanics over time as a guide to degassing rate (figure 4.2) is shown in figure 5.13 and compared to values of RCO_2 obtained from using spreading rate. Results show that for both formulations the effects of degassing on CO_2 are greatest in the Mesozoic and Cenozoic (past 250 million years). The results for

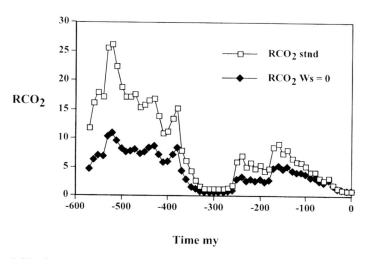

Time my

Figure 5.11. Comparison sensitivity plot of RCO_2 versus time compared to the GEOCARB III standard curve for the situation of holding Ws = 0. Ws is the coefficient that relates the rate of weathering to changes in temperature due to changes in solar radiation. Ws = 0 means no solar evolution over time, with irradiation held constant at present levels.

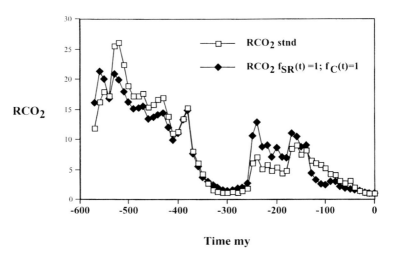

Figure 5.12. Comparison sensitivity plot of RCO_2 versus time for the situation of $f_{SR}(t) =1$ and $f_C(t) = 1$ for all time compared to the GEOCARB III standard curve. This shows the effect on CO_2 of holding global CO_2 degassing to be the same as at present.

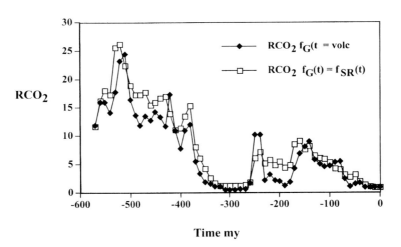

Figure 5.13. Comparison sensitivity plot of RCO_2 versus time for the situation of $f_G(t) = f_V(t)$ for all time compared to the GEOCARB III standard curve based on $f_G(t) = f_{SR}(t)$. The variable $f_G(t)$ is the rate of global CO_2 degassing at a past time (t) divided by that for the present. The variable $f_V(t)$ is the value for $f_G(t)$ calculated from the abundance of volcanic rocks over time (Ronov, 1993). The variable $f_{SR}(t)$ is the value of $f_G(t)$ calculated from changes in seafloor spreading rate over time.

all three formulations, no change in degassing, use of spreading rate, and use of volcanic abundance, result in curves of RCO_2 that are crudely similar. This means that the effects of parameters other than degassing, as discussed above, are more critical to the overall Phanerozoic history of atmospheric CO_2. Either further improvements in degassing formulation are necessary, or degassing has had a lesser effect on Phanerozoic CO_2 evolution than is commonly believed by most workers (e.g., Berner et al., 1983; Tajika, 1998; Wallmann, 2001; Schrag, 2002). Considering our present state of knowledge concerning degassing (see chapter 4), it is probable that the truth lies somewhere between these two extremes. At any rate, I suspect that the GEOCARB spreading rate degassing formulation probably needs to be changed.

For the sake of completeness, I present an example for the results of models of short-term perturbations to the long-term carbon cycle. A characteristic of all these models are asymmetric curves of CO_2 and other output variables versus time. A sudden change is followed by a gradual return to steady state as the inputted carbon perturbation is diluted and carried through the long-term cycle (Caldeira and Rampino, 1990; Kump, 1991; Dickens et al., 1997; Dessert, 2001; Beerling and Berner, 2002; Beerling et al., 2002; Berner, 2002). The example chosen here is that presented by Berner (2002) to explain a large negative excursion of $\delta^{13}C$ of sedimentary carbonates across the Permian-Triassic boundary. Two explanations that have been offered for the isotope excursion are CO_2 degassing associated with volcanic eruption from Siberian basalts at that time and release of CH_4 from methane hydrates, which are oxidized rapidly to CO_2 in the oceans or atmosphere. In the model, carbon dioxide with $\delta^{13}C = -6‰$ is released from the mantle over 200,000 years (from 2.5×10^6 km^3 of basalt), or methane with $\delta^{13}C = -65‰$ is released from methane hydrates over 20,000 years and rapidly oxidized to CO_2. Both releases follow a Gaussian distribution with a maximum release rate halfway through the release period (see figure 5.14). The results for $\delta^{13}C$, based on reasonable estimates of the time-integrated release of 11000 GtC of volcanic CO_2 or 4200 GtC of CH_4, are shown in figure 5.15. The drop of about 7‰ in $\delta^{13}C$ from CH_4 degassing matches observations (e.g., Twitchett et al., 2001), but such a large drop could not be attained by appealing to volcanic degassing (volcanic CO_2 in figure 5.15). This is because volcanic carbon, derived from the mantle, is not sufficiently isotopically light. A drop of 7‰ would require impossibly high fluxes of volcanic CO_2.

Proxy Methods

A variety of independent tests of the modeling predictions as to levels of Phanerozoic atmospheric CO_2 are available. Methods include determining (1) the $\delta^{13}C$ of carbonates in paleosols; (2) the stomatal density

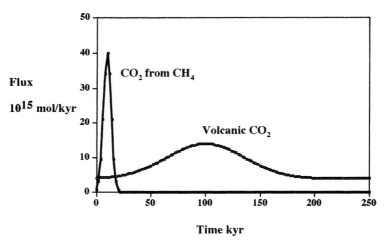

Figure 5.14. Input function for volcanic degassing of CO_2 and for CO_2 formed by the oxidation of CH_4 from methane hydrate decomposition at the Permian–Triassic boundary. (After Berner, 2002.)

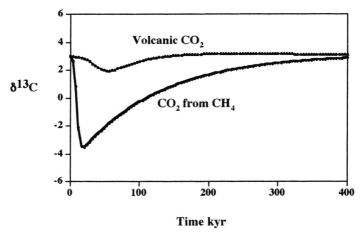

Figure 5.15. Calculated plots of $\delta^{13}C$ of $CaCO_3$ versus time at the Permian–Triassic boundary. The upper curve is for volcanic degassing associated with the eruption of Siberian basalts. The lower curve is for the input of CO_2 from the oxidation of CH_4 derived from methane hydrates. The return of the curves to the initial conditions represent dilution of the inputs as the long-term carbon cycle returns to a steady state. Actual measured values of $\delta^{13}C$ show changes of over –7‰ during this same time span thereby supporting the methane source hypothesis. (Data from Berner, 2002.)

of fossil leaves; (3) the degree of fractionation of carbon isotopes of specific compounds secreted by phytoplankton and preserved in sedimentary rocks; and (4) the boron isotopic composition of marine carbonate fossils. Each of the methods has its own problems (Royer et al., 2001a), but if certain precautions are taken, they provide reasonable estimates of ancient CO_2 levels.

Paleosol Carbon Isotopes

The paleosol method (Cerling, 1991) rests on the assumption that the carbon in pedogenic carbonate is precipitated in isotopic equilibrium with dissolved inorganic carbon in soil waters and with soil CO_2. The isotopic composition of soil CO_2 is the result of mixing respired CO_2, with the $\delta^{13}C$ value of soil organic matter, and atmospheric CO_2. The relative proportions of respired CO_2 and atmospheric CO_2 determine the $\delta^{13}C$ of the soil gas and, if exchange equilibrium is attained, also of soil carbonate. Determining the $\delta^{13}C$ of organic matter and carbonate in a given paleosol, estimating the original partial pressure of CO_2 at depth in the paleosol, and estimating the $\delta^{13}C$ of atmospheric CO_2 allows a mass balance calculation of the partial pressure of atmospheric CO_2 at the time that the paleosol formed. The mass-balance equations used for this purpose is

$$\partial C / \partial t = D \partial^2 C / \partial z^2 + \phi(z) \qquad (5.11)$$

where

C = concentration of CO_2 in the soil gas
z = depth in soil
$\phi(z)$ = production rate of respiratory CO_2
D = diffusion coefficient for gaseous CO_2.

A steady state is assumed, where diffusion balances CO_2 production, and equation (5.11) is set equal to 0. The resulting expression for C is

$$C = C_a + [\phi(0)z^{*2}/D]\,[1 - \exp(-z/z^*)] \qquad (5.11)$$

where C_a is CO_2 concentration in air, and z^* is characteristic CO_2 production depth where $\partial C / \partial z$ approaches 0.

From a similar mass-balance expression for $\delta^{13}C$, and substituting appropriate values of key parameters, Cerling (1991) obtains:

$$C_a = S(z)[(\delta_s - 1.044\delta_\phi - 4.4)/(\delta_a - \delta_s)] \qquad (5.12)$$

where $S(z) = [\phi(0)z^{*2}/D]\,[1 - \exp(-z/z^*)]$, $\delta = \delta^{13}C‰$, and the subscripts s, ϕ, and a refer to soil, respired, and atmospheric CO_2, respectively.

By substituting values for the three δ values, one can solve for C_a, the concentration of CO_2 in the atmosphere. The value for δ_s is calculated for equilibrium with the measured value of $\delta^{13}C$ for $CaCO_3$ formed in the soil. The value for δ_ϕ is assumed to be the same as the measured (or often assumed) $\delta^{13}C$ for soil organic matter associated with the carbonate. The value for δ_a is calculated from the value for δ_ϕ or from the mean $\delta^{13}C$ value for the oceans as registered by marine carbonates deposited at the same approximate time as the paleosol. This assumes that the isotopic fractionation between soil organic carbon and atmospheric CO_2 or between dissolved inorganic carbon in seawater (δ_{oc}) and atmospheric CO_2 each exhibit a constant value. For the present earth these differences are

$$\delta_a - \delta_{oc} \cong -7\%o \tag{5.13}$$

$$\delta_a - \delta_\phi \cong 18\%o \tag{5.14}$$

To solve for C_a, the paleosol $\delta^{13}C$ method is forced to make several assumptions, the most important of which concerns the value of S(z), which can be considered as the mean concentration of CO_2 added to the soil via respiration. The value of S(z) can be estimated by looking at total concentrations of CO_2 in modern $CaCO_3$-containing soils. Pedogenic carbonates only form under rather dry climates with less than 80 cm of mean annual precipitation (Royer, 1999), and such soils have S(z) concentrations ranging from 3000 to 10000 ppm (Royer et al., 2001). It is this range of ignorance that provides the major portion of error attached to reported results using this method. Other possible errors include failure to estimate the proper temperature at the time of formation of the paleosol (which affects the isotopic fractionation between CO_2 and $CaCO_3$) and changes in the values of fractionations shown in equations (5.13) and (5.14). In many paleosols the value of $\delta^{13}C$ for organic matter is not measured, and δ_ϕ has to be calculated from δ_{oc} via both equations (5.13) and (5.14). Another more serious problem is that the carbonate analyzed must be true pedogenic carbonate and not derived from diagenetic recrystallization, especially of marine carbonates. Thus, the method requires field experience in studying and identifying paleosols (e.g., see Retallack, 1990).

A variant of the paleosol pedogenic carbonate method has been adapted by Yapp and Poths (1992, 1996). They measure the $\delta^{13}C$ value of trace amounts of carbon included within the structure of the mineral goethite ($HFeO_2$). The reasoning behind calculating concentrations of atmospheric CO_2 is the same as in the carbonate method, except that the concentration of carbon contained within the iron oxide is used as a direct measure of soil CO_2 partial pressure. In this way the necessary assumption of a wide range of values for soil CO_2 can be avoided. However, organic matter is rarely present in pedogenic goethites, and estimates of $\delta^{13}C$ of the organic matter from an extrapolation procedure leads to the considerable error with this method. A more important problem is the rarity of appropriate unaltered soil goethite, as compared to soil $CaCO_3$.

Results for paleosol estimates of atmospheric CO_2 are shown in figure 5.16 and compared to the standard (best estimate) curve from the GEOCARB III model for the entire Phanerozoic. The solid line represents a five-point running average (Royer et al., 2001a) based on hundreds of analyses of Carboniferous to recent paleosols (Cerling, 1991; Sinha and Stott, 1994; Andrews et al., 1995; Ekart et al.,1999, Lee, 1999; Lee and Hisada, 1999; Ghosh et al., 2001). Also included are determinations by Mora et al. (1996) and Yapp and Poths (1992, 1996) for Devonian and earlier paleosols. In general there is qualitative to semiquantitative agreement between the theoretical GEOCARB modeling and the paleosol CO_2 estimates, with high values during the early Paleozoic (before 380 Ma), a large drop during the Devonian and Carboniferous (380–320 Ma), a rise to high values in the Mesozoic (250–65 Ma), and a gradual overall decline during the Cretaceous and Cenozoic (140–0 Ma). The very high palesol values during the mid-Mesozoic around 180 Ma are biased by two extremely high ($RCO_2 \cong 20$) Jurassic values of Yapp and Poths (1992, 1996), and I believe these values are excessive.

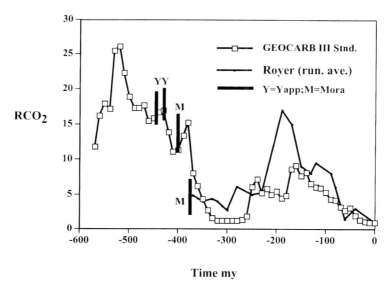

Figure 5.16. Comparison of RCO_2 determined by the paleosol carbon isotope method to GEOCARB III modeling. Y= Yapp and Poths (1992, 1996); Mora = Mora et al. (1996); Royer = running 5-point average of all post-Devonian paleosol results summarized by Royer et al. (2001a) from a variety of studies (see text).

Because of special interest in the rise of large vascular land plants during the Devonian and their effect on atmospheric CO_2, an expanded plot is shown for this time period (figure 5.17). Data for paleosols are from Yapp and Poths (1992, 1996), Mora et al. (1996), Mora and Driese (1999), and Cox et al. (2001). Also shown are paleo-CO_2 estimates via the stomatal index method (McElwain and Chaloner, 1995, 1996; discussed next). Both the two kinds of proxies and the GEOCARB modeling show excellent agreement, and this further emphasizes the importance of the rise of large vascular land plants during the Devonian (380–350 Ma) as a major effect on atmospheric CO_2.

Stomatal Index

The density of stomata (gas exchange openings) on leaves can vary with the CO_2 level in the atmosphere (Woodward, 1987). The density also varies with changes in water availability, called "water stress." A simple way to eliminate the water stress effect is to normalize the density of stomata to the number of epidermal cells, which corrects for changes in cell size (Salisbury, 1927). The result is called stomatal index. Besides water stress, stomatal index is insensitive, also to illumination and temperature (Beerling, 1999). The selective sensitivity to CO_2 makes stomatal index a potentially powerful tool for deducing ancient CO_2

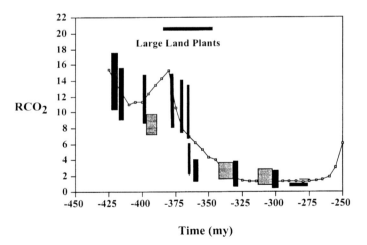

Time (my)

Figure 5.17. Plots of RCO_2 versus time for the paleosol carbon isotope method (vertical bars) and the stomatal index method (square boxes) compared to the standard GEOCARB III curve. Data for the paleosol carbon isotope method from Yapp and Poths (1992, 1996), Mora et al. (1996), Mora and Driese (1999), and Cox et al. (2001). Stomatal index data from McElwain and Chaloner (1995, 1996). All methods indicate that a large drop in CO_2 occurred at the same time as the rise of large vascular land plants.

levels from the study of fossil leaves, but it is necessary to choose leaves for study that are either conspecific with modern representatives or that have nearest living ecological and morphological equivalents. In this way the CO_2 response of the stomata can be calibrated. Also, fossil leaves need to be collected that were derived from trees that were tall or in sparse stands so that they did not sample excess CO_2 that builds up at shallow levels in forests, and the trees must not have been at high elevations because stomata respond to the partial pressure of CO_2 and not to concentration (Royer et al., 2001). At sea level concentration and partial pressure are identical.

McElwain and Chaloner (1995, 1996) and McElwain (1998) have used the nearest-living-equivalent approach for the study of the stomatal index of fossil leaves ranging from the Devonian period to the present. They calculate paleo-CO_2 levels by the "stomatal ratio" method, which assumes that the ratio of stomatal index for two different times is directly proportional to the inverse of their CO_2 levels. They calibrate their method by choosing a paleo-CO_2 level that shows rough agreement between different methods ($RCO_2 = 1$ at 300 Ma; see figure 5.17). This method is useful in showing semiquantitative effects, but it is not calibrated against a known level of CO_2. A plot of results for RCO_2 derived via the stomatal ratio method is compared to the standard curve for GEOCARB III in figure 5.18, and the overall trend with time is similar to that based on modeling. Also, the large drop in CO_2 based on modeling and paleosol results (figure 5.17) is corroborated by the stomatal ratio method.

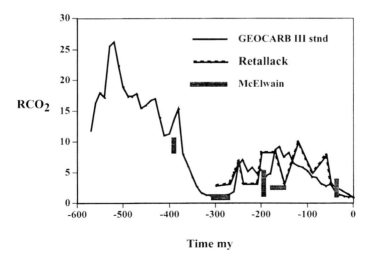

Figure 5.18. Plots of RCO_2 versus time derived by the stomatal index method compared to the standard GEOCARB III curve. Stomatal results from McElwain (1998) and Retallack (2001) (5-point running average).

Quantitative calibration of stomatal index (SI) against CO_2 change has been done via both laboratory experiments and the history of CO_2 over the past 150 years as recorded in ice cores and actual atmospheric measurements. Both approaches have been used by Royer et al. (2001b) to calibrate *Ginkgo biloba* and *Metasequoia glyptostroboides* leaves. Values of SI were measured on samples from herbarium collections that had been collected at different times over the past 150 years at atmospheric CO_2 levels ranging from 270 ppm to 360 ppm. These results were combined with *Ginkgo* and *Metasequoia* experiments conducted at elevated CO_2 levels. An excellent linear negative correlation between SI and CO_2 concentration was found for the ginkgo and Metasequoia leaves between 270 and 360 ppm, but at higher CO_2 levels, based on laboratory experiments, the response diminished and approached constant SI values. This shows that the SI method may have an upper CO_2 limit, at least on the time scale (few years) of laboratory experiments.

Retallack (2001) extrapolated from the SI versus CO_2 experimental calibrations of Beerling et al. (1998) for a single ginkgo species to obtain values of RCO_2 over the past 300 million years based on the study of other species. Although the extrapolation is not theoretically justified, and CO_2 response can vary greatly between species (Royer et al., 2001), Retallack's results, in terms of 5-point moving averages, show overall semiquantitative agreement with those of GEOCARB modeling and the stomatal ratio results of McElwain (1998; figure 5.18).

The results of Royer et al. (2001) and other well-calibrated SI studies based on single species (van der Burgh et al., 1993; Kurschner, 1997) are summarized for the Tertiary below in figure 5.20. Results for the Eocene disagree with those based on the plankton fractionation and boron isotope methods (described below), but there is general agreement for the Miocene.

Plankton Carbon Isotope Fractionation

The fractionation of carbon isotopes by marine phytoplankton is a function of the concentration of dissolved CO_2 in seawater (Rau et al., 1989; Hinga et al., 1994). If surface waters can be assumed to be in exchange equilibrium between dissolved CO_2 and atmospheric CO_2, then the fractionation could be used as a measure of the level of atmospheric CO_2. This observation has led to the suggestion that the difference between the $\delta^{13}C$ of coexisting organic matter and $CaCO_3$ in sedimentary rocks might be a measure of ancient atmospheric CO_2 concentration (Jasper and Hayes, 1990; Hollander and McKenzie, 1991; Freeman and Hayes, 1992). However, the organic matter in a sedimentary rock represents a mixture of phytoplankton-derived material, material that has been processed along the marine food chain and any terrestrially derived material added to the ocean by rivers. Because the $\delta^{13}C$ of these sources are different, determining $\delta^{13}C$ of the bulk organic matter would not reflect the value of the original photosynthate. To avoid this problem, a method has been developed for analyzing only organic compounds that are produced by and confined to phytoplankton. Using this method, ancient atmospheric CO_2 levels have been deduced (Freeman and Hayes, 1992) and are plotted in figure 5.19. These results, as for the results of the paleosol and stomatal index methods, show a similar trend to that predicted by the GEOCARB III model, that of decreasing CO_2 over the Mesozoic and Cenozoic.

Unfortunately, the plankton fractionation method is confronted with the problem that there are additional factors in the marine environment, other than CO_2, that affect carbon isotope fractionation during photosynthesis. This introduces errors not shown by the error margins of the data in figure 5.19. Additional factors include plankton growth rate (Fry and Wainright, 1991) and cell geometry (Popp et al., 1998). Detailed experimental study of these effects has enabled Bidigare et al. (1997) to correct for these factors. These investigators found that growth rate is a direct function of the concentration of dissolved phosphate in seawater. To normalize for constant growth rate, Pagani et al. (1999) applied the fractionation method to sediments that were deposited under oligotrophic conditions similar to those at present where the phosphate correction can be made. Pagani et al. obtained values for Miocene paleo-CO_2 based on measurements of $\delta^{13}C$ of alkenones and carbonate microfossils, and results are shown in figure 5.20 in combination with the estimates of other methods for Tertiary CO_2.

The difference between the $\delta^{13}C$ of bulk organic matter and calcium carbonate ($\Delta^{13}C$), as documented by Hayes et al. (1999), has been used

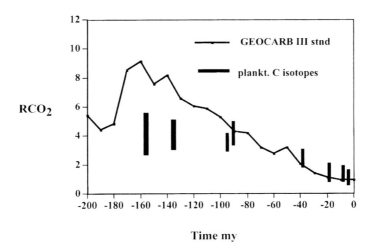

Figure 5.19. Comparison of results for RCO_2 versus time based on the plankton carbon isotope method (Freeman and Hayes, 1992) with the standard GEOCARB III curve.

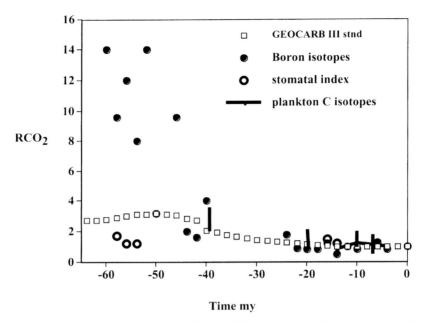

Figure 5.20. Comparison of results for Tertiary RCO_2 versus time. Boron isotope results from Pearrson and Palmer (1999, 2000); stomatal index results from Royer et al. (2001), van der Burgh (1993), and Kurschner (1997); plankton carbon isotope results from Freeman and Hayes (1992; vertical bars) and Pagani et al. (1999; curved bar). The succession of small unfilled boxes represents results for GEOCARB III modeling.

to calculate the value of atmospheric CO_2 concentration over Phanerozoic time (Rothman, 2002). However, as stated above, bulk organic matter represents a mixture of terrestrial and marine-derived material, and the calibrations based on the study of marine plankton do not apply generally to bulk material. Also, it has been shown that fractionation of carbon isotopes by plant-derived organic matter is not a simple function of CO_2, but rather a strong function of atmospheric O_2 (Beerling et al., 2002), which varies with time (see chapter 6). The intermixture of large amounts of terrestrially derived material with $\delta^{13}C$ values affected by variations in O_2 is especially true of the Permo-Carboniferous, when a large proportion of global organic burial was derived from land plants (chapter 3). These considerations indicate that the simple use of $\Delta^{13}C$ to derive CO_2 values over the Phanerozoic (Rothman, 2002) is an inappropriate approach to the problem of deducing paleo-CO_2.

Boron Isotopes in Carbonates

Dissolved boron in the oceans is present primarily as $B(OH)_3$ and $B(OH)_4^-$, and these two species differ in the ratio $^{11}B/^{10}B$. Field observations and experimental studies have suggested that the uptake of boron into biogenic calcium carbonate records the isotopic composition of $B(OH)_4^-$ with little isotopic fractionation (Hemming and Hanson, 1992; Sanyal et al., 1996). Because the relative proportions of the two dissolved boron species vary with pH, and the degree of isotopic fractionation between the two species is known, the $^{11}B/^{10}B$ of $B(OH)_4^-$ also varies with pH (Hemming and Hanson, 1992; Sanyal et al., 1996). If the boron incorporated into calcareous organisms faithfully records the isotopic composition of $B(OH)_4^-$ in seawater, it is possible to calculate the value of pH for ancient oceans by determining the $^{11}B/^{10}B$ of fossil organisms (Hemming and Hanson, 1992; Sanyal et al., 1995, 1996; Palmer et al., 1998; Pearson and Palmer, 1999, 2000). From paleo-pH, making certain assumptions about the concentration of dissolved inorganic carbon (DIC) in seawater, it is possible to calculate paleo-atmospheric CO_2 from paleo-pH. This is the basis for the boron isotopic method for estimating paleo-CO_2.

There are problems, however, with this method. First, the fractionation of boron isotopes is a function of temperature and this requires an estimate of the water temperature where the calcareous organism actually lived. (This factor can be helpful in deducing the depth distribution of paleo-pH when applied to planktonic foraminifera that are known to live at different temperatures at different water depths; Pearson and Palmer, 1999). Second, it is assumed that only $B(OH)_4^-$ is taken up in the carbonate which, on the basis of the crystal structure of calcite and aragonite, is difficult to explain. Third, it is assumed that there is no fractionation (Hemming and Hanson, 1992; Palmer et al., 1998) or a known constant degree of fractionation for a given organism (Sanyal et al., 1996) during uptake of $B(OH)_4^-$. This has been challenged by Vengosh et al. (1991) and by M. Pagani (personal communication). Fourth, the isotopic composition of total boron in seawater is assumed constant over time, which is justified for periods much shorter than the mean residence time of boron in seawater (18 million years), but not for the Tertiary (Lemarchand et al., 2000). Finally, there is the problem of converting paleo-pH to paleo-CO_2. This requires knowledge of the DIC of paleo-seawater, which is not known and must be estimated. As CO_2 changes over millions of years, the concentration of DIC must also change (Caldeira and Berner, 1999).

Pearson and Palmer (1999, 2000) have estimated CO_2 levels ranging from the Paleocene to the Pliocene, using the boron isotopic method applied to planktonic foaminfera. Their results are also shown in figure 5.20, along with results for GEOCARB modeling, the stomatal index method, and the planktonic carbon isotope fractionation method. Strong disagreement between the boron isotope method and the other methods are obvious for the Eocene (55–34 Ma), but not for the Miocene (24–5 Ma). Overall, as pointed out earlier, there is general agreement for all methods of decreasing CO_2 during the past 65 million years.

Summary: CO_2 and Climate

Both theoretical modeling and proxy estimates point to similar trends for CO_2 over Phanerozoic time. High levels of CO_2 in the early Paleozoic are followed by a large drop to low levels during the Permo-Carboniferous. A primary reason for this drop, based on GEOCARB modeling, and the models of Mackenzie et al. (2003) and Bergman et al. (2003), was the rise of large vascular land plants, which accelerated the weathering of Ca and Mg silicates and later led to the burial of increased quantities of organic matter in sediments. At the end of the Permian, possibly at the very end, the level of atmospheric CO_2 rose to high values during the early Triassic. After this, CO_2 remained high in the Mesozoic and began a gradual decline, punctuated by short- and medium-term excursions, which extended into and through the Cenozoic, reaching the rather low

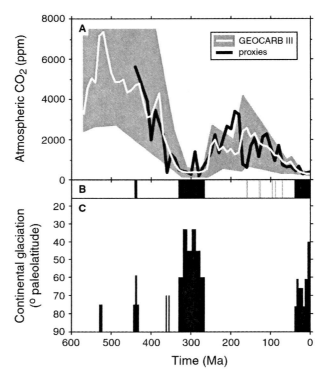

Figure 5.21. CO_2 and climate (Modified from Royer et al., 2004).
A: Comparison of model predictions (GEOCARB III) and proxy estimates of CO_2. Shaded area represents range of error for model calculations. All proxy results are averaged within successive ten million years time slots. B: Intervals of glacial (black) or cool climates (light gray). C: Latitudinal distribution of direct glacial evidence (tillites, striated bedrock, etc.). (After Royer et al., 2004.)

level, compared to most of the Phanerozoic, found at present. Contributing to this drop was increased mountain uplift in the Cenozoic and a continual increase in solar radiation throughout the Phanerozoic. The solar increase exerted a major influence on weathering rate, helping to bring about an overall long-term decline in CO_2.

Changes in CO_2 over the Phanerozoic correlate rather well with changes in paleoclimate. Times of minimal CO_2 coincide with the two most widespread and long-lasting glaciations of the Phanerozoic, that during the Permo-Carboniferous (330–270 Ma) and that of the past 30 million years (Crowley and Berner, 2001; Royer et al., 2004). This is illustrated in figure 5.21. Also, periods of unusual warmth at high latitudes, such as occurred during much of the Mesozoic (250–65 Ma) correlate with periods of elevated CO_2. Together these observations give support to the greenhouse theory of climate change on a Phanerozoic time scale (Royer et al., 2004). Short-lived glaciations and cooling episodes, such as occurred during the late Ordovician at 440 Ma, can be explained by other causes but may also involve short-term changes in atmospheric CO_2.

There are problems with all of the methods of CO_2 estimation, and they have been pointed out in this and (for modeling) in previous chapters. The GEOCARB standard curve is not intended to be used as an accurate CO_2 measure (as is sometimes mistakenly done), but rather as a suggestion of how CO_2 has changed over the Phanerozoic. New advances in the various fields that contribute to the modeling (e.g., volcanology, biogeochemistry of weathering, tectonics, and erosion) will undoubtedly cause modifications to the GEOCARB curve, and to those of Mackenzie et al. (2003) and Bergman et al. (2003). However, the overall qualitative trend described above, I believe, will withstand the test of time. It has now stood for more than 14 years.

6

Atmospheric O_2 over Phanerozoic Time

The chemical reactions that affect atmospheric O_2 on a multimillion-year time scale involve the most abundant elements in the earth's crust that undergo oxidation and reduction. This includes carbon, sulfur, and iron. (Other redox elements, such as manganese, are not abundant enough to have an appreciable effect on O_2.) Iron is the most abundant of the three, but it plays only a minor role in O_2 control (Holland, 1978). This is because during oxidation the change between Fe^{+2} and Fe^{+3} involves the uptake of only one-quarter of an O_2 molecule, whereas the oxidation of sulfide to sulfate involves two O_2 molecules, and the oxidation of reduced carbon, including organic matter and methane, involves between one and two O_2 molecules. The same stoichiometry applies to reduction of the three elements. Because iron is not sufficiently abundant enough to counterbalance its low relative O_2 consumption/release, the iron cycle is omitted in most discussions of controls on atmospheric oxygen. In contrast, the sulfur cycle, although subsidiary to the carbon cycle as to its effect on atmospheric O_2, is nevertheless non-negligible and must be included in any discussion of the evolution of atmospheric O_2.

In this chapter the methods and results of modeling the long-term carbon and sulfur cycles are presented in terms of calculations of past levels of atmospheric oxygen. The modeling results are then compared with independent, indirect evidence of changes in O_2 based on paleobiological observations and experimental studies that simulate the response of forest fires to changes in the levels of O_2. Because the sulfur cycle is not discussed anywhere else in this book, it is briefly presented first.

The Long-Term Sulfur Cycle and Atmospheric O_2

The long-term sulfur cycle is depicted as a panorama in figure 6.1. Sulfate is added to the oceans, via rivers, originating from the oxidative weathering of pyrite (FeS_2) and the dissolution of calcium sulfate minerals (gypsum and anhydrite) on the continents. Volcanic, metamorphic/hydrothermal, and diagenetic reactions add reduced sulfur to the oceans and atmosphere where it is oxidized to sulfate. Sulfur is removed from the oceans mainly via formation of sedimentary pyrite and calcium sulfate. Removal also occurs at mid-ocean rises (not shown) via hydrothermal pyrite and $CaSO_4$ formation, but the pyrite formation is small relative to that forming in sediments, and the $CaSO_4$ is mostly subsequently redissolved (see discussion below).

The major overall reactions in the long-term global sulfur cycle that affect atmospheric O_2 are

$$15O_2 + 4FeS_2 + 8H_2O \rightarrow 2Fe_2O_3 + 8SO_4^{-2} + 16H^+ \qquad (6.1)$$

$$2Fe_2O_3 + 8SO_4^{-2} + 16H^+ \rightarrow 15O_2 + 4FeS_2 + 8H_2O \qquad (6.2)$$

(Pyrite, FeS_2 is shown as a generalization for all sulfide sulfur, including that in organic sulfides.) It is notable that the essence of these reactions, which are complex combinations of several intermediate reactions as shown below, were deduced by Ebelmen (1845)[1] 120 years before they were independently reformulated by Garrels and Perry (1974), Garrels et al. (1976), and Holland (1978). Reaction (6.1) represents mainly the oxidative weathering of pyrite on the continents. It also represents the summation of several steps involving the thermal breakdown of pyrite at depth (including the mantle) followed by the oxidation by atmospheric O_2 of reduced sulfur emitted to the atmosphere and oceans via volcanoes and hot springs. For example, one possible pathway (involving later oxidation of all reduced products at the earth surface) is:

$$4FeS_2 + 20H_2O + 4SiO_2 \rightarrow 8SO_2 + 20H_2 + 4FeSiO_3 \qquad (6.3)$$

$$8SO_2 + 4O_2 + 8H_2O \rightarrow 8SO_4^{-2} + 16H^+ \qquad (6.4)$$

---

1. Ebelmen was one of the first to use the new chemical symbolism of Berzelius, which has been used ever since, but his reactions were written in terms of 30 O instead of 15 O_2 because the oxygen molecule had not yet been discovered.

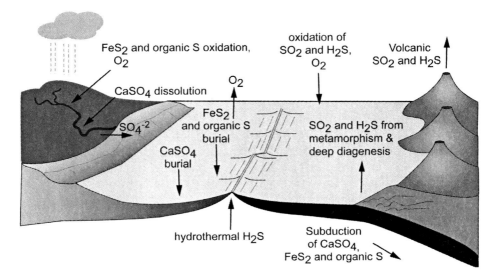

Figure 6.1. The long-term sulfur cycle. Downward pointing arrows associated with O_2 signify O_2 consumption; upward pointing arrows signify O_2 production.

$$4FeSiO_3 + O_2 \rightarrow 2Fe_2O_3 + 4SiO_2 \tag{6.5}$$

$$20H_2 + 10O_2 \rightarrow 20H_2O \tag{6.6}$$

Overall, the sum of reactions (6.3) to (6.6) is the same as reaction (6.1). Reaction (6.2) is also an overall reaction involving several steps:

$$16CO_2 + 16H_2O \rightarrow 16CH_2O + 16O_2 \tag{6.7}$$

$$16CH_2O + 8SO_4^{-2} \rightarrow 8H_2S + 16HCO_3^- \tag{6.8}$$

$$2Fe_2O_3 + 8H_2S + O_2 \rightarrow 4FeS_2 + 8H_2O \tag{6.9}$$

$$16H^+ + 16HCO_3^- \rightarrow 16CO_2 + 16H_2O \tag{6.10}$$

The sum of these reactions is the same as reaction (6.2). Reaction (6.7) represents burial of photosynthetic carbon in sediments, reaction (6.8) is bacterial sulfate reduction, reaction (6.9) is sedimentary pyrite formation, and reaction (6.10) is the neutralization of bicarbonate. For further details on the overall process of sedimentary pyrite formation, consult Morse et al. (1987).

In comparison to these rather complex reactions affecting O_2, the remaining reactions of the long-term sulfur cycle, that involve calcium sulfate minerals, are simple:

$$CaSO_4 \rightarrow Ca^{++} + SO_4^{-2} \tag{6.11}$$

$$Ca^{++} + SO_4^{-2} \rightarrow CaSO_4 \tag{6.12}$$

Reaction (6.11) represents the dissolution of gypsum and anhydrite in sedimentary rocks during continental weathering, whereas reaction (6.12) represents the precipitation and burial of these same minerals in sediments deposited in evaporite basins. Since evaporitic sedimentary rocks occur sporadically throughout the geologic record, it is likely that removal fluxes from the ocean via $CaSO_4$ formation are often unbalanced relative to addition fluxes from weathering (Berner and Berner, 1996), leading to non–steady-state levels of sulfate in seawater on a million-year time scale.

Tectonic processes involving sulfur, as major controls on atmospheric O$_2$, have been emphasized by Walker (1986) and by Hansen and Wallmann (2003). This includes emission of reduced sulfur-containing gases during volcanism and hydrothermal reactions at mid-ocean spreading centers (figure 6.1). According to Hansen and Wallmann, reactions at the spreading centers are complex and involve $CaSO_4$ precipitation and later dissolution, reduction of seawater sulfate by reaction with Fe^{+2} in basalt with the formation of hydrothermal pyrite, and oxidation of primary sulfides and H_2S derived from the mantle. The overall net effect of these pocesses is the consumption of atmospheric O$_2$.

The quantitative importance of sulfur reactions occurring at mid-ocean spreading centers has been questioned by Berner et al. (2003). First of all, the fluxes calculated by Hansen and Wallmann (2003) depend strongly on calculations of the rate of water flow through the mid-ocean rises and how it has changed with time, which is a controversial subject (Edmond et al., 1979; Morton and Sleep, 1985; Kadko, 1996; Rowley, 2002). Second, the most important process, hydrothermal $CaSO_4$ precipitation, is essentially balanced by later dissolution of the $CaSO_4$. Third, the sulfur isotopic composition of H_2S emitted at ridge hydrothermal vents indicates that more than 80% of the sulfur is derived from the mantle and not from the reduction of seawater sulfate. Finally, oxidation of Fe^{+2} minerals in basalt by the reduction of seawater sulfate results in an irreversible atmospheric oxygen drop over long times (Petsch, 1999).

Model Calculations

Calculation of the change in atmospheric oxygen with time is straightforward based on reactions (1.7), (1.8), (6.1), and (6.2):

$$d[O_2]/dt = F_{bg} - F_{wg} + (15/8)(F_{bp} - F_{wp}) \tag{6.13}$$

where

[O_2] = mass of oxygen in the atmosphere

F_{bg} = rate of burial of organic carbon in sediments

F_{wg} = rate of oxidative weathering of organic carbon plus rate of oxidation of reduced carbon-containing gases released via diagenesis, metamorphism, and volcanism

F_{bp} = rate of pyrite (plus organic) sulfur burial in sediments

F_{wp} = rate of oxidative weathering of pyrite (and organic) sulfur plus rate of oxidation of reduced sulfur-containing gases released via diagenesis, metamorphism, and volcanism.

The 15/8 term refers to the stoichiometry of reactions (6.1) and (6.2). This equation shows that calculation of O_2 over time requires a knowledge of the various fluxes: F_{bg}, F_{wg}, F_{bp}, F_{wp}. These fluxes have been estimated basically by three approaches: (1) the "rock abundance" method, which involves quantifying the amounts of organic carbon and pyrite sulfur originally buried in Phanerozoic sedimentary rocks; (2) the isotope mass balance method, which consists of calculations based on the carbon and sulfur isotopic composition of the oceans, as recorded in sedimentary rocks; and (3) the a priori method, which is based on theoretical models for the burial and oxidation of carbon and sulfur.

The Rock Abundance Method

The rock abundance method (Berner and Canfield, 1989), for F_{bg} and F_{bp}, is based on original global sedimentation rates of terrigenous sediments (sandstones and shales) and their organic carbon and pyrite sulfur contents. (Most sedimentary organic matter and pyrite is found in shales.) For this purpose terrigenous deposition over time is divided into three major categories (Ronov, 1976): coal basin sediments, noncoal continental deposits (mainly redbeds), and marine sediments. This leads to the expressions:

$$F_{bg} = (f_{mar}C_{mar} + f_{cb}C_{cb} + f_{rb}C_{rb})\,F_t \qquad (6.14)$$

$$F_{bp} = [O_2(0)/O_2](f_{mar}S_{mar} + f_{cb}S_{cb} + f_{rb}S_{rb})\,F_t \qquad (6.15)$$

where

f_{mar}, f_{cb}, f_{rb} = fraction of total terrigenous sedimentation deposited as marine sediments, coal basin sediments, and redbeds

C_{mar}, C_{cb}, C_{rb}, S_{mar}, S_{cb}, S_{rb} = mean organic carbon and pyrite (+ organic) sulfur content of marine sediments, coal basin sediments, and redbeds respectively

F_t = total global rate of terrigenous sedimentation

$O_2(0)/O_2$ = ratio of oxygen mass at present to that at some prior time.

Changes in the relative proportions f_{mar}, f_{cb}, f_{rb} of each type bring about changes in global rates of burial of organic C and pyrite S (the terms for redbeds are dropped because they contain negligible organic C and pyrite S). The term $O_2(0)/O_2$ is added to include the inverse effect of O₂ levels on the burial of pyrite (Berner and Raiswell, 1983). The fraction of terrigenous sediments as coal basin sediments and as redbeds is shown in figure 6.2 as a function of time. Because the average organic carbon content of coal basin sediments (mainly dispersed at subeconomic levels) are about 3–5 times higher than that for average marine terrigenous sediments and because redbeds contain essentially no organic carbon or pyrite sulfur, these plots are a qualitative measure of O₂ production rate versus time. High proportions of coal basin sediments mean high O₂ production, and high proportions of redbeds means low O₂ production.

To complete their calculations of O₂ versus time, Berner and Canfield (1989) assumed that the weathering of organic C and pyrite S was a direct function of physical erosion rate. This is in keeping with the find-

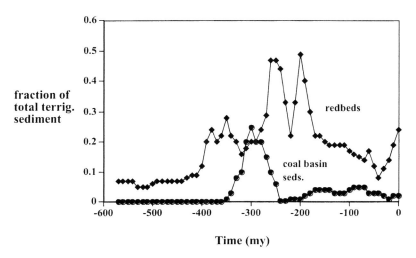

Figure 6.2. Plots versus time of the fraction of terrigenous sediments (sandstones and shales) present as coal basin sediments and as other terrestrial deposits (mainly redbeds). Data from Ronov (1976).

ings on present-day organic matter and pyrite weathering discussed in chapter 3. Because global erosion equals global sedimentation, this provides a strong negative feedback against variations in O_2 due to variations in organic C burial accompanying variations in total global sedimentation. Also, to provide additional negative feedback against excessive O_2 variation, Berner and Canfield introduced the concept of rapid recycling. In rapid recycling, the organic C and pyrite S in younger sediments are assumed to weather faster than in older ones that have become buried and sheltered from the atmosphere. As a simple first-order approach, sediments were divided into rapidly weathering "young" sediments with a mean age of 100 million years and all other slowly weathering "old" sediments with greater ages.

The Isotope Mass Balance Method

The isotope mass balance method for determining the values of F_{bg}, F_{wg}, F_{bp}, and F_{wp} is based on variations over Phanerozoic time in $^{13}C/$ ^{12}C and $^{34}S/^{32}S$ of the oceans as recorded by the isotopic composition of $CaCO_3$ and sulfate-containing minerals (mainly $CaSO_4$) in sedimentary rocks (Veizer et al., 1999; Strauss, 1999). As discussed in chapter 3, changes in oceanic $^{13}C/^{12}C$ (figure 3.3), expressed as $\delta^{13}C$, largely reflect changes in the burial flux of organic matter in sediments. Changes in oceanic $^{34}S/^{32}S$ largely reflect changes in the burial flux of pyrite in sediments because bacterial sulfate reduction to H_2S (reaction 6.8) brings about the depletion of ^{34}S in the resulting sedimentary pyrite. Removal of carbon as carbonates and sulfur as sulfates from seawater involves little isotope fractionation. Thus, increased burial of organic matter and pyrite results in greater removal of the lighter isotopes, causing the oceans to become enriched in ^{13}C and ^{34}S. There is little fractionation of carbon and sulfur isotopes during the weathering or thermal decomposition of organic matter, pyrite, carbonates, and sulfates. Thus, the isotopic composition of input to the oceans and atmosphere reflects the relative importance of the weathering and thermal decomposition of isotopically light sulfide and organic C versus that of isotopically heavy carbonate and sulfate.

The isotope mass balance modeling method is illustrated in figure 6.3. Mass balance calculations for total carbon, total sulfur, ^{13}C, and ^{34}S are used to determine rates of weathering, thermal decomposition, and burial of organic matter, pyrite, carbonates, and sulfates and their changes over Phanerozoic time. The mass balance expressions (equations 1.10. and 1.11.) for carbon and ^{13}C were presented in chapters 1 and 5. The analogous expressions for sulfur are

$$dM_s/dt = F_{ws} + F_{wp} + F_{ms} + F_{mp} - F_{bs} - F_{bp} \qquad (6.16)$$

$$d(\delta_s M_s)/dt = \delta_{ws}F_{ws} + \delta_{wp}F_{wp} + \delta_{ms}F_{ms} + \delta_{mp}F_{mp} - \delta_{bs}F_{bs} - \delta_{bp}F_{bp} \quad (6.17)$$

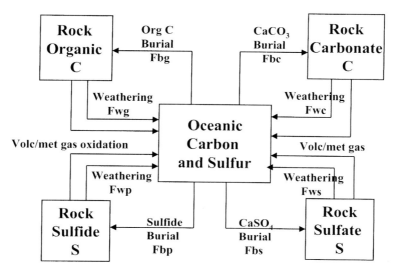

Figure 6.3. Isotope mass balance model for the carbon and sulfur cycles as they affect atmospheric oxygen.

where

M_s = mass of sulfur in the oceans (the total mass in the atmosphere, biosphere, and soils is by comparison negligible)
F_{wp} = sulfate flux from oxidative weathering of pyrite
F_{ws} = sulfate flux from weathering of Ca sulfates
F_{mp} = sulfur degassing flux for pyrite from volcanism, metamorphism, and diagenesis
F_{ms} = sulfur degassing flux for Ca sulfates from volcanism, metamorphism, and diagenesis
F_{bs} = burial flux of Ca sulfates in sediments
F_{bp} = burial flux of pyrite in sediments
$\delta = [(^{34}S/^{32}S)/(^{34}S/^{32}S)\text{stnd} - 1]\ 1000.$

It is assumed that the value of δ_{bs} is the same as δs for seawater and that the difference between δ_{bs} and δ_{bp}, representing the fractionation of sulfur isotope during bacterial sulfate reduction, is constant or varies with time.

The various fluxes based on the use of expressions analogous to equations (1.10), (1.11), (6.16), and (6.17) have been calculated by a number of workers (Holland, 1978; Veizer et al., 1980; Berner and Raiswell, 1983; Garrels and Lerman, 1984; Francois and Gerard, 1986; Kump and Garrels, 1986; Walker, 1986; Berner, 1987, 2001; Kump, 1989; Lasaga, 1989; Carpenter and Lohmann, 1997; Petsch and Berner, 1998; Bergman et al., 2003; Hansen and Wallmann, 2003; Mackenzie et al., 2003). However, most of these studies have focused on only one element, either carbon or sulfur, and not on the evolution of atmospheric O_2. In fact, some studies (e.g., Garrels and Lerman, 1984), have assumed constant O_2. In all cases the isotope mass balance modeling is guided by the isotopic composition of seawater as recorded by carbonates or sulfates in sedimentary rocks. The change in seawater isotopic composition with time represents the balance between input fluxes of ^{13}C or ^{34}S from weathering and thermal degassing and outputs via burial in sediments. Calculations were performed in two ways. The first method starts with assumed initial values and readjusts them until correct present-day values are obtained at t = 0. The other method starts with present-day values and then calculates the fluxes backwards in time.

In the modeling studies cited above, except for the recent ones of Hansen and Wallmann (2003), Bergman et al. (2003), and Mackenzie et al. (2003), weathering and thermal degassing are lumped together as "weathering," and total crustal carbon and sulfur are assumed to be conserved. Also, the pre-2003 studies assume that the various (f) factors discussed throughout this book are equal to 1. This means that weathering rate is simply proportional to the mass of each reservoir shown in figure 6.3. For further simplification, in a number of studies (e.g., Berner, 2001) steady state is assumed so that dM_c/dt, dM_s/dt, $d(\delta_c M_c)/dt$ and $d(\delta_s M_s)/dt$ are all set equal to 0. For carbon this assumption is very reasonable (e.g., see chapter 1), and inequalities between weathering and burial of organic matter result in reciprocal inequalities between weathering and burial of $CaCO_3$. Total sulfur is also conserved by reciprocal behavior of pyrite and $CaSO_4$, but this is less justifiable than in the case of carbon because a modest portion of total sulfur can be stored as dissolved sulfate in the oceans.

Any large difference in the net flux between oxidized and reduced reservoirs of carbon or sulfur cannot be maintained over millions of years, or fluctuations in atmospheric O_2 would occur that are either physically impossible (negative masses) or so high that they are incompatible with the persistence of life on earth. One way to eliminate such impermissible fluctuations is to have any imbalances in the carbon cycle closely balanced by corresponding imbalances in the sulfur cycle (Lasaga, 1989). This would be reflected by inverse plots of $\delta^{13}C$ and $\delta^{34}S$ versus time, a situation that is approximated by existing isotopic data (Veizer et al., 1980). However, use of actual carbon and sulfur isotopic data does not remove this problem. The simple mass balance models, when applied to both C and S isotopic data, result in excessive fluctuations in O_2 because of the great sensitivity of O_2 to the isotopic data (Berner, 1987; Lasaga, 1989; Berner, 2001).

The only way that excessive variations of O_2 can be avoided when modeling both carbon and sulfur isotopes is to introduce negative feedback. One way to do this is to have pyrite and organic matter weathering rate increase with increasing atmospheric O_2. However, as pointed out in chapter 3, studies of present-day organic matter weathering have not verified this assumption. It is commonly believed (e.g., Holland, 1978; Berner, 2001) that the rate-limiting step is exposure of organic matter to the atmosphere by physical erosion and not the reaction of the organic matter with O_2. In addition, use of a direct proportionality between the rate of organic weathering and O_2 concentration does not remove excessive O_2 variation but, in fact, introduces new destabilizing positive feedback (Lasaga, 1989).

To remedy the problem of excessive O_2 variation, two mechanisms that provide negative feedback, but that are not normally considered as feedback processes, have been used. These are rapid recycling, as applied to rock abundance modeling discussed above, and O_2-dependent carbon and sulfur isotope fractionation (Berner, 2001). Rapid recycling alone results in some reduction of the excessive fluctuations, but there are still physically impossible values for O_2 (figure 6.4). In contrast, oxygen-dependent isotope fractionation for carbon and sulfur, when added to rapid recycling, does result in reasonable values of O_2 over time (figure 6.4).

There is independent evidence that O_2-dependent carbon isotope fractionation actually occurs. Laboratory plant-growth experiments show that $^{13}C/^{12}C$ fractionation during the growth of a variety of land plants increases with increases in the ratio O_2/CO_2, an observation expected from

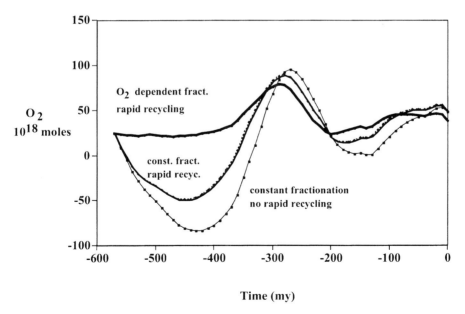

Time (my)

Figure 6.4. Plots of the mass of atmospheric O_2 versus time calculated via C and S isotope mass balance modeling. Realistic values of O_2 mass are obtained if both rapid recycling and O_2-dependent isotope fractionation are included in the modeling.

photosynthetic theory (Beerling et al., 2002). Also, O_2-dependent fractionation calculated from modeling agrees with measurements of the difference in $\delta^{13}C$ between coexisting organic matter and carbonates over most of the Phanerozoic (Hayes et al., 1999; Berner, 2001). (Further discussion of plants and O_2-dependent carbon isotope fractionation is presented later in this chapter.) For sulfur it is reasonable to assume that $^{34}S/^{32}S$ fractionation during pyrite formation should also be a function of O_2. There is increased fractionation of sulfur isotopes between seawater sulfate and pyrite sulfur when, during the process of pyrite formation at shallow sediment depths, there is reoxidation of sulfur due to bioturbation (Canfield and Teske, 1996). With higher atmospheric O_2, one would expect greater oxidation of sulfide accompanying the addition of dissolved O_2 into sulfide-rich sediments by the bioturbating organisms.

The A Priori Method

A third approach, not dependent directly on either rock abundance or carbon and sulfur isotopic data, has been used for calculating changes in atmospheric O_2 over time. Most of the models consider the cycles of the nutrients, P and N, and how they bring about negative feedback stabilization of O_2 level. (Such feedbacks, as they affect marine organic matter burial, are illustrated in figure 3.1 of chapter 3.) The model of Hansen and Wallmann (2003) for the past 150 million years uses the a priori approach and formulates the effects of various processes on burial, weathering, and degassing fluxes including the use of modifying nondimensional (f) parameters like those discussed in this book. They calculate not only O_2 but CO_2 and the composition of seawater over time. In their model the burial of organic carbon is assumed to be proportional to the rate of supply of phosphorus to the oceans, which arises from the weathering of organic matter, silicates, and carbonate and is inversely proportional to atmospheric O_2. In terms of the nomenclature of the present book (see table 5.1):

$$F_{bg} = [(F_{wg} + F_{wsi} + F_{wc})/(F_{wg}(0) + F_{wsi}(0) \\ + F_{wc}(0))] \times [O_2(0)/O_2]^m F_{bg}(0) \qquad (6.18)$$

where O_2 is the mass of O_2 in the atmosphere, m is an arbitrary parameter (0 to 1), and (0) refers to quaternary average values. For pyrite burial Hansen and Wallmann assume

$$F_{bp} = r_{mar} \, r_{an} \, (S/C) \, F_{bg} \qquad (6.19)$$

where

r_{mar} = fraction of total organic burial occurring in marine sediments
r_{an} = fraction of marine organic burial occurring in anoxic sediments
S/C = pyrite sulfur/organic carbon burial ratio (assumed constant).

The oxidative weathering of organic matter and pyrite is assumed to follow

$$F_{wg} = f_R(t)f_A(t)F_{wg}(0) \qquad (6.20)$$

$$F_{wp} = f_R(t)f_A(t)F_{wp}(0) \qquad (6.21)$$

Hansen and Wallmann (2003) separate true weathering from oxidation of reduced gases produced by thermal decomposition during volcanism, metamorphism, and hydrothermal reactions occuring at mid-ocean ridges. Expressions for the thermal reactions are numerous and are not repeated here.

Hansen and Wallman (2003) use carbon and sulfur isotopic data, but only to guide their modeling. They calculate values of δ^{13}C and δ^{34}S over time and then adjust parameters (such as m in equation 6.18) to get the best fit with measured values. For both isotopes this results in excellent agreement with measurements, giving strong backing to the model assumptions. One problem with applying their model to longer periods is the assumption of a constant C/S ratio for organic matter burial. During the Paleozoic there were great changes in C/S due to shifting deposition of organic matter between anoxic (euxinic) basins (low C/S), terrestrial coal swamps (very high C/S), and normal (non-euxinic) marine sediments (intermediate C/S) (Berner and Raiswell, 1983; see also figure 3.2).

Another a priori model is that of Lenton (2001). This model is actually only an addition of modifying factors to the Phanerozoic rock abundance model of Berner and Canfield (2001). He ignores pyrite burial and weathering and multiplies the right side of equation (6.14) by a factor representing the effects of plants on the weathering release of phosphate. The idea is that phosphorus is the limiting nutrient for organic carbon production and burial and that more P release by weathering brings about greater organic matter burial (see figure 3.1). Plant abundance, and therefore the rate of P weathering, is constrained by the effect of O$_2$ on plant productivity and fire frequency. This provides negative feedback to constrain excessive O$_2$ variation, an idea also suggested by Kump (1988) for P release by forest fires. Unfortunately, Lenton's model is not physically correct because the original model of Berner and Canfield (equation 6.14) is for actual measured organic C in sediments, and the calculated F_{bg} value cannot be arbitrarily varied with modifying factors.

Bergman et al. (2003) have presented an a priori comprehensive model for the entire Phanerozoic that tracks both atmospheric O$_2$ and CO$_2$ and the composition of seawater with time. Factors considered by the modeling include the cycles of C, S, P, and N, interactive marine and terrestrial biota, changing solar insolation, metamorphic and volcanic degassing, tectonic uplift, apportioning carbonate burial between shallow and deep marine sediments, land plant evolution, and plant-assisted weathering. In several aspects the model resembles the GEOCARB model, but it is more complex in considering the atmosphere separately from the oceans and tracking seawater composition. A similar GEOCARB-like approach has been presented by Mackenzie et al. (2003) in a model for Phanerozoic O_2, CO_2, sediment chemistry, and seawater chemistry. Their model is the most complex to date, and they add a host of new processes. They consider the cycles of C, S, Al, Si, Ti, Ca, Mg, Cl, Fe, K, Na, P, and Si and divide their system into five reservoirs: shallow and deep cratonic carbonate and silicate rocks and sediments, seawater, the atmosphere, oceanic sediments and basalts, and the shallow mantle. They also consider the following processes: continental and seafloor weathering of silicates and carbonates, sedimentary dolomite formation, net ecosystem productivity, seawater-basalt exchange, precipitation and diagenesis of chemical sediments (including formation of new silicates by "reverse weathering"), redox reactions involving C, S, and Fe, and subduction-decarbonation reactions.

Model Results

Results of rock abundance modeling (Berner and Canfield, 1989) and isotope mass balance modeling (Berner, 2001) for atmospheric oxygen over Phanerozoic time are compared in figure 6.5. The isotope modeling, designated here as RROD, is based on the carbon isotopic data of Veizer et al. (1999) (with a minor positive excursion during the Silurian ignored; see figure 3.3), the sulfur isotopic data of Strauss (1999), and rapid recycling (RR) and O_2-dependent (OD) C and S isotope fractionation (Berner, 2001). There is excellent agreement between the results of the two approaches. This is partly due to adjusting the O_2 dependence of carbon and sulfur isotopic fractionation to obtain a best fit to the rock abundance results. However, the resulting fractionations for carbon isotopes $\Delta^{13}C$ are in rough agreement (except for the past 20 million years) with the data for the measured difference between $\delta^{13}C$ values for coexisting carbonates and organic matter in sediments (Hayes et al., 1999). This is shown in figure 6.6. Note that values of the adjustment parameters in the RROD model (J and n; see Berner, 2001, for details) are the same for both the O_2 calculation (figure 6.5) and the $\Delta^{13}C$ calculation (figure 6.6). Agreement of the plots in each figure gives some credence to the results of the RROD isotope mass balance method. Furthermore, measurements of carbon isotopic fractionations for plant fossils (discussed below) provide further agreement with the theoretically calculated Permo-Carboniferous values of $\Delta^{13}C$ centered around 300 Ma. (Much of global organic burial during the Permo-Carboniferous derived from land plants; Berner and Raiswell, 1983.)

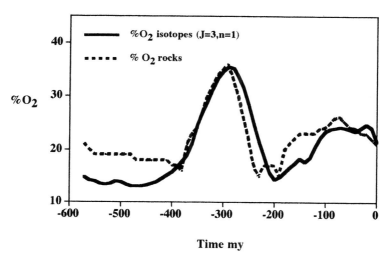

Figure 6.5. Plots of % O_2 versus time based on rock abundance modeling (Berner and Canfield, 1989) and RROD isotope mass balance modeling with rapid recycling and O_2-dependent C and S isotope fractionation. The adjustable parameters J and n refer to the isotope dependence of fractionation for C and S isotopes, respectively (Berner, 2001).

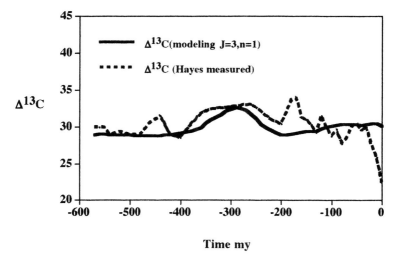

Figure 6.6. Comparison of carbon isotope fractionation $\Delta^{13}C$ over time for the RROD model with measurements of $\Delta^{13}C$ by Hayes et al. (1999) on Phanerozoic carbonates and organic matter.

The most notable feature of figure 6.5 is the maximum in O_2 concentration extending from 375Ma to 250Ma (Devonian–Permian). This is the result of an increased rate of production of O_2 by an increased rate of burial of organic matter during this period. The increased burial rate is indicated by both the abundance of organic matter in sediments and by the carbon isotopic composition of the oceans. High values of oceanic $\delta^{13}C$ during this period (figure 3.3) indicate increased removal of light carbon from seawater via photosynthesis, resulting in increased burial of the produced organic matter in sediments. Increased burial of organic matter on land (e.g., in coal basin sediments) would also be recorded by heavy carbon isotopic values in seawater because of rapid isotopic exchange of the atmosphere with the oceans. The increased burial was brought about mostly by the rise of large woody land plants (i.e., trees). Woody plants contain lignin, which is relatively resistant to microbial decomposition, and deposition of lignin must have led to the addition of organic matter to terrestrial sediments and extra organic matter to marine sediments after transport there by rivers. (For further discussion of organic burial, see chapter 3.)

The evolution of atmospheric O_2 during the past 150 Ma has been calculated by Hansen and Wallmann (2003), using an a priori method, and is compared in figure 6.7 to the RROD isotope model. There is amazingly good agreement considering the use by Hansen and Wallmann of f factors, as they affect weathering, which is not done in the RROD modeling. The disagreement between the two studies for the period 100–50 Ma

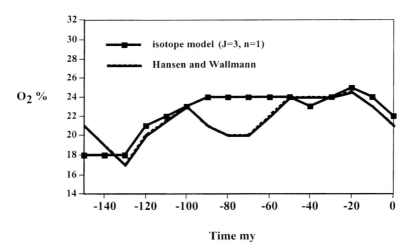

Figure 6.7. Comparison of results for % O_2 for the past 150 million years between the model of Hanson and Wallmann (2003) and RROD isotope mass balance modeling (Berner, 2001).

can be explained by the lack of consideration of these factors by the simple isotope model.

The results for O_2 level over Phanerozoic time, calculated via the a priori model of Bergman et al. (2003), are compared to the results of RROD modeling in figure 6.8. Both models show a large rise in percent O_2 during the mid-Paleozoic, attaining a maximum near the Permian-Carboniferous boundary. However, the Bergman model shows another equally high maximum in the Cretaceous not obtained by RROD isotope modeling.

Independent Indicators of Phanerozoic O_2

The theoretical calculations of the evolution of atmospheric O_2 over Phanerozoic time can be checked against geological observations. Because the metabolism of plants and animals are sensitive to levels of O_2 in the environment, one can look at the paleontological and paleobotanical record for suggestions of possible changes in atmospheric O_2 and limits on maximum and minimum O_2 concentrations. To do this on a quantitative basis, experiments on the response of modern organisms to changes in O_2 are necessary. Combining modern experiments with observations of physiologically dependent features of fossil organisms leads to the underexplored field of paleophysiology. (A pioneering and largely forgotten paper on this subject is that of McAlester, 1970.) In

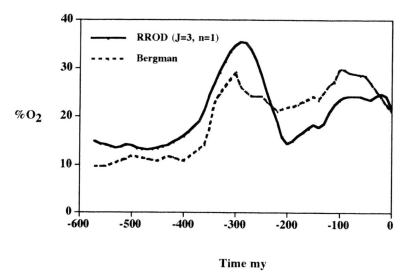

Figure 6.8. Comparison of calculated values of atmospheric O_2 over Phanerozoic time by RROD isotope mass balance modeling and the a-priori modeling of Bergman et al. (2003).

addition to paleophysiology, limits on O_2 variation also can be made by studying experimentally the effect of O_2 concentrations on fire behavior and comparing results with the evidence for paleofires, such as charcoal occurrences and the fire resistance of ancient plants.

Plants and Oxygen

Before the Miocene, plants of the geological past almost all utilized the C-3 photosynthetic pathway that involves the competition of O_2 and CO_2 for sites on the key enzyme, rubisco. Carbon dioxide is taken up by rubisco during photosyhthesis, and O_2 is taken up during photorespiration. Fractionation of carbon isotopes is affected by the relative rates of photosynthesis and photorespiration. At high O_2/CO_2 more O_2 is taken up on rubisco and less CO_2. As a result, the CO_2 builds up within the cell, and, according to photosynthetic theory (Farquhar and Wong, 1982), this should lead to a greater fractionation of carbon isotopes:

$$\Delta^{13}C = a + (b - a)\ c_i/c_a \qquad (6.22)$$

where

c_i = intercellular CO_2 concentration
c_a = CO_2 concentration in the external air
$a = \Delta^{13}C$ that occurs during diffusion through the stomata (4.4‰)
$b = \Delta^{13}C$ for the fixation of CO_2 by Rubisco (27–29‰).

An increase of c_i, due to CO_2 accumulation as a result of being outcompeted by rising O_2, should, therefore, lead to a rise in $\Delta^{13}C$.

Laboratory experiments with growing plants verify this prediction (Berry et al., 1972; Beerling et al., 2002). The measured fractionation $\Delta^{13}C$ between atmospheric CO_2 and plant carbon has been found to be proportional to the ratio O_2/CO_2 (figure 6.9). This is exactly what would be predicted from equation (6.22) and the above discussion. This observation suggests that the fractionation of carbon isotopes, as recorded by plant fossils, might be used to deduce ancient levels of atmospheric O_2, providing that an independent estimate of atmospheric CO_2 could be made. However, even without any CO_2 data, it is still possible to use qualitatively the idea that higher concentrations of atmospheric O_2 would lead to greater fractionation. To test this idea, measurements of $\delta^{13}C$ for 41 plant fossils ranging in age from Devonian to Cretaceous were performed (Beerling et al., 2002). To be able to determine $\Delta^{13}C$, values of $\delta^{13}C$ for atmospheric CO_2, of the same age as each plant fossil, were calculated from data for surface water oceanic carbon, as recorded by carbonate fossils (Veizer et al., 1999). The $\delta^{13}C$ value for atmospheric CO_2 should exhibit a constant difference with oceanic $\delta^{13}C$ because of rapid air–water exchange and equilibration (Cerling, 1991; see equation 5.13).

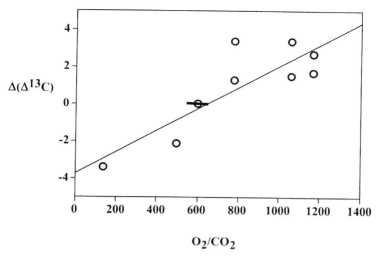

$\Delta(\Delta^{13}C)$

O_2/CO_2

Figure 6.9. Plot of relative carbon isotopic fractionation versus O_2/CO_2 of the atmosphere based on experimental plant growth experiments. The parameter $\Delta(\Delta^{13}C)$ represents the deviation of fractionation of each experiment from a paired control held at ambient levels of 21% O_2 and the same CO_2 level. The point with a horizontal bar through it represents O_2/CO_2 for 21% O_2 and 0.030–0.036% CO_2. Values of O_2 used for each paired experiment ranged from 4% to 35% and from 0.030% to 0.045% for CO_2. Species examined were *Phaseolous vulgaris*, *Sinapis alba*, *Ravunculus repens*, and *Macrozamia communis*. The data of Berry et al. (1972) for *Atriplex patula* $\Delta(\Delta^{13}C)$ at 4% and 21% O_2 are included in the figure. (Data from Beerling et al., 2002.)

Results for $\Delta^{13}C$, based on plant fossil measurements for the Devonian–Triassic are shown in figure 6.10 and compared with predicted values from RROD isotope mass balance modeling. The good agreement gives further support to the hypothesis that levels of atmospheric O_2/CO_2 were unusually high at this time and that these high levels brought about an increase in photosynthetic carbon isotope fractionation. The recent study of a much larger number of plant fossils and terrestrial organic matter by Peters-Kottig et al. (2003) also found increased fractionation during the late Paleozoic and ascribed it to elevated O_2/CO_2. These studies illustrate how the study of plant physiology supplements geochemical modeling in examining the history of atmospheric O_2.

Elevated levels of atmospheric O_2 are deleterious to plants. This is because of enhanced photorespiration, as discussed above, and because the production of toxic OH radicals is increased by a higher level of respired O_2 (Fridovich, 1975; Lane, 2002). This suggests an upper limit to the concentration of O_2 in the atmosphere. As a test of the conclusions of Berner and Canfield (1989) that maximal Permo-Carboniferous concentrations of O_2 were 30–35%, Beerling et al. (1998) conducted

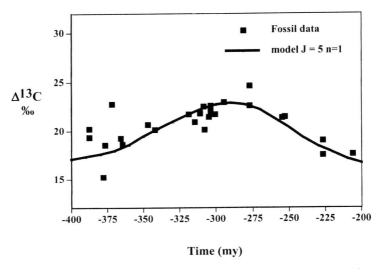

Figure 6.10. Plots of carbon isotope fractionation based on measurements of Devonian–Triassic plant fossils compared to predictions by the RROD isotope mass balance model. (Data from Beerling et al., 2002.)

plant growth laboratory experiments at these levels and at ambient (360 ppm) CO_2. The results were that at 35% O_2, plant net primary production decreased by about 20%. This shows that at such a high level of O_2, plant growth is inhibited but that plants are not killed, at least not on a month-to-year time scale. Furthermore, calculations by Beerling and Berner (2000) indicate that raising the concentration of atmospheric CO_2, at the Permo-Carboniferous O_2 maximum, from 300 ppm to 450 ppm, a value permitted by the errors of GEOCARB modeling, results in a rate of net primary production that is essentially the same as under present-day, preindustrial conditions (300 ppm CO_2, 21% O_2). This is not unexpected because the preindustrial present-day ratio of O_2/CO_2 (210,000 ppm/300 ppm) is reasonably close to that hypothesized for the Permo-Carboniferous O_2 maximum (350,000 ppm/450 ppm).

Animals and Oxygen

All multicellular animal life is dependent on the uptake of oxygen during respiration. Thus, any change in the level of O_2 might be expected to affect animal physiology. Physiology includes such things as thermal balance, respiratory gas exchange, and animal flight. Animals with lungs have an advantage over those without lungs because they have the ability to change breathing rates and to evolve bigger or smaller lungs according to the availability of O_2. However, in animals that metabolize by passive diffusion of O_2 into their bodies, respiration should be readily affected by changes in atmospheric O$_2$. This is especially true of insects (Vogel, 1994; Dudley, 2000). Larger insects, with a lower surface area-to-mass ratio, should respire more slowly at a fixed level of O$_2$. However, if O$_2$ concentration were to rise, for a constant body size and shape, this would allow increased respiration for the larger sizes. Proof that the metabolism of dragonflies is O$_2$-limited during flight has been demonstrated by Harrison and Lighton (1998). This suggests that an elevated atmospheric O$_2$ concentration could allow for an increase in the size of this insect.

In the fossil record, giant insects, including dragonflies (Graham et al., 1995; Dudley, 1998; Lane, 2002), are confined to the Permo-Carboniferous time of hypothesized high O$_2$ level. The dragonfly fossils reach wing spans up to 80 cm, and their thoraxes are so large that they could not obtain enough O$_2$, under present-day levels of 21% O$_2$, to be able to fly (Lane, 2002). Along with dragonflies, there are unusually large amphibians, mayflies, millipedes, hexapods, and arachnids confined to this same time span (Dudley, 1998), and these organisms also metabolize by passive diffusion. Thus, animal fossils provide further evidence for the hypothesized high O$_2$ concentrations during the Permo-Carboniferous.

Further proof that insects will adapt to an increase in O$_2$ concentration by increasing in size is shown by the experiments of R. Dudley (see Berner et al., 2003). Dudley exposed five generations of *Drosophila* fruit flies to hyperbaric (1.1 atm) conditions such that the partial pressure of O$_2$ was 0.23 atm (equivalent to 23% O$_2$ at one atmosphere total pressure). Each successive adult generation was found to become bigger, and the mean size eventually leveled off at a constant value, indicating acclimitization to elevated O$_2$. Restoring the sixth generation to ambient conditions allowed for the exclusion of intragenerational phenotypic plasticity. Results shown in table 6.1 indicate a statistically significant increase in mean size after the six generations for both males and females. This is in comparison to a normobaric (21% O$_2$, 1.0 atm total pressure) control run simultaneously. It is unfortunate that the experi-

Table 6.1. Results of experiments after subjecting five generations of *Drosophila melanogaster* to P_{O_2} = 0.23 atm at a total pressure of 1.1 atm.

P_{O_2} (atm)	Gender	No. of samples	Body mass (mg) ($\pm$1 SD)
0.21	Males	30	0.64 $\pm$ 0.07
0.23	Males	36	0.73 $\pm$ 0.12
0.21	Females	32	1.07 $\pm$ 0.19
0.23	Females	31	1.18 $\pm$ 0.24

Experiments of R. Dudley from Berner et al. (2003).

ment was not continued and the fifth generation of fruit flies subjected to further increases of O_2 partial pressure to see if additional increases in mean size would ensue. In a separate experiment, subjecting a single generation of *Drosophila* to a sudden large increase to 40% O_2 did not result in a consistent increase in mean size (Frazier et al., 2001), probably due to the inability of the organisms to adjust this quickly to the adverse effects of hyperoxia such as excessive OH radical production.

These exploratory studies suggest a need for further experimental work on the growth and metabolism of insects and other diffusion-respiring animals under elevated O_2. For this purpose an initial study on amphibian physiology and growth under elevated O_2 is underway at Yale University. Results may help explain the presence of giant amphibians at the time of hypothesized maximal atmospheric oxygen.

Fires and Oxygen

The frequency of forest fires, all other factors being held constant, should vary with the level of atmospheric oxygen because fire is sensitive to oxygen concentration. The quantitative effect of varying O_2 on fire ignition and spread has been studied extensively in the laboratory (e.g., Tewarson, 2000; Babrauskas, 2002), but only on commercial materials. Little attention has been paid to the effects of varying O_2 on the burning of forest fuels. One such study examined the ignition of paper strips at varying O_2 levels and moisture contents (Watson, 1978). On the basis of Watson's results, it was concluded that forests would burn excessively at concentrations of 25% O_2 or higher even when saturated with moisture (Watson et al., 1978). Because plant fossils indicate that forests have persisted over the past 375 million years, Watson et al. (1978) concluded that O_2 levels have never exceeded 25% over this time.

The conclusions of the widely cited Watson et al. (1978) paper (e.g., see Lenton, 2001) have been criticized by Robinson (1989). Robinson notes that (1) paper is thermally thin and not representative of naturally thick plant material. This thinness maximizes exposure to oxygen. (2) Charring was not considered. Paper is very low in lignin and does not char as readily as natural lignin-rich plant materials (table 6.2). Charring acts as a protective shield to protect plants against further burning (Nelson, 2001). (3) Moisture contents far less than found in natural vegetation were studied. (4) Translation of O_2 results to fuel moisture equivalents and an index of probability of ignition was done using a table that has since been replaced by the U.S. Forest Service because it proved unreliable (Deeming et al., 1977).

Watson's (1978) experiments, and those of Rashbash and Langford (1968), deduced a lower limit for forest burning of about 15% O_2. In other words, forests could not have been ignited if the level of O_2 in the geological past had ever dropped below about 15%. Because forest fires are documented by the occurrence of charcoal (as fusain) in the fossil record

Table 6.2. Charring propensity for natural fuels.

Material	Char fraction (%)
Filter paper (cellulose)	4.2
Poplar wood (low lignin)	14.1
Chamise foliage, live[a]	39.4
Bracken tree fern fronds, cured[a]	42.0
Ponderosa pine bark[a]	44.3

Data from Berner et al. (2003).

[a]Fire adapted.

ever since the Devonian (Cope and Chaloner, 1980; Chaloner, 1989), this indicates that O_2 has not dropped below this value (Chaloner, 1989, suggests a lower limit of 13%). Although discernment of this lower level needs further investigation, it is in approximate agreement with recent burning experiments (Wildman et al., 2004).

Some data based on recent, more realistic, burning experiments (Wildman et al., 2004) are shown in table 6.3. These were conducted over a range of O_2 concentrations of 8–35% (O_2/N_2 gas mixtures) and moisture contents of 0–61% dry weight for pine wood and 0–190% for pine needles. In the "wood" experiments, a 4-cm thick layer of diagonally criss-crossed 10 cm × 1 cm x 1 cm square wooden dowels were placed on a 75-cm long path; for pine needles a constant thickness of

Table 6.3. Results of burning experiments with pine wood dowels and pine needles.

Moisture (%)	O_2 content (%)					
	12%	16%	21%	28%	31%	35%
Pine wood						
0–2	Yes	Yes	Yes	Yes	Yes	Yes
12	—	No	Yes	Yes	—	Yes
23	—	—	No	No	Yes	Yes
61	—	—	No	No	No	No
Pine needles						
0–2	No	Yes	Yes	Yes	Yes	Yes
12	—	—	Yes	—	—	Yes
23	—	—	Yes	—	—	Yes
61	—	—	—	—	—	No
193	—	—	—	—	—	No

"Yes" means that the fire spread completely over a 75-cm long track. "No" means that the initial fire went out within a few centimeters of the starting point. Moisture is for a dry weight basis. A dash means no experiment. (After Wildman et al., 2004).

4 cm was used over the same path and a fire with dry dowels or dry needles was started at one end. The spreading of the fire down the path was tracked with thermocouples and the spreading rate thereby obtained. Results show that at low moisture levels (0–2% water) the wooden dowels all burned readily between 12% and 35% O_2, whereas at 61% water (by dry weight) the dowels did not burn at any O_2 level up to and including 35% O_2 (table 6.3). The pine needles with low (0–2%) moisture burned between 16% and 35% O_2, but at 61% water and 35% O_2 they did not burn and presumably would not burn at lower O_2 levels at this moisture level.

On the basis of fire spreading rate, Wildman et al. (2004) concluded that the sensitivity to changes in moisture is much larger than the sensitivity to changes in oxygen level. The lack of burning at 35% O_2, for both pine wood and pine needles containing 61% H_2O, refutes the idea that forest materials will burn above 25% O_2 regardless of moisture content (Watson et al., 1978). (The water content of live tree trunks normally exceeds 100% by dry weight and saturated pine needles contain 190% water.) Starting the fire with dry wood or needles crudely simulates a brush fire that spreads to moister fuels, so these experiments with natural materials are more realistic than igniting paper strips. Based on the burning experiments of Wildman et al. (2004), it is quite possible that oxygen levels of the Permo-Carboniferous exceeded 30% without the destruction of all terrestrial life.

The results of the burning experiments of Watson (1978) and Wildman et al. (2004) both point to increased burning at higher O_2 levels, suggesting increased fire frequency during the Permo-Carboniferous high-O_2 period and natural selection in favor of fire-resistant plants at that time. The slow rise of O_2 through the Devonian and Carboniferous predicted by modeling (Berner and Canfield, 1989; Berner, 2001) would allow enough time for the evolution of widespread fire defenses. The common occurrence of thick, barklike corky layers on the outside of Carboniferous plants, such as lycopsids and *Calamites* (Jeffry, 1925, cited in Komrek, 1972) were probably developed as a defense against fire. Also, besides bark structure and composition, the spatial organization, energy partitioning, and reproductive strategies of Carboniferous ecosystems appear to be consistent with those of forests subject to severe fire regimes (Robinson 1989, 1991). Furthermore, present-day species that are probable relics of Carboniferous plants are associated with areas of frequent forest fires (Komrek, 1972).

Late Carboniferous coal beds probably arose from raised megathermal peat bogs analogous to the present-day swamps of Indonesia and Malaysia (Robinson, 1989). However, the Carboniferous coals are unusually rich in fusain, fossil charcoal, compared to their relatively charcoal-free modern analogues. As charcoal is a product of fires, this suggests a greater swamp fire frequency during the Late Carboniferous than at present, due to presumably higher O_2 levels. Also, promotion of char formation is an effective way to create fire resistance (Nelson, 2001). This suggests that the high fusain (fossil charcoal) content of the Carboniferous coals could indicate an abundance of plants with increased fire resistance. In addition, the formation and burial of charcoal, which is resistant to biological oxidation, is a positive feedback mechanism for the enhancement of atmospheric O_2 level (Berner et al., 2003).

The reflectivity of charcoal is directly proportional to the temperature at which the charcoal formed (Scott, 2000). Jones and Chaloner (1991) have shown that the cell wall morphology of fusainized woods also reflects the temperature of formation. If fire temperature can be correlated with O_2 level, then fusain reflectivity and/or cell wall morphology might be a guide to ancient atmospheric O_2 concentration. However, many factors may affect fire temperature other than atmospheric O_2 level, and, in fact, it is difficult to reconstruct fire regimes from the nature of charcoal even in modern sediments (Clark et al., 1997). More work on the properties of fossil charcoal is needed before anything more definitive can be stated concerning the usefulness of charcoal as a paleo-O_2 indicator.

Summary

Atmospheric O_2 concentration has likely varied considerably over Phanerozoic time. Theoretical modeling, based on the abundance of organic carbon and pyrite sulfur in sedimentary rocks, and the carbon and sulfur isotopic composition of carbonates and sulfates in sedimentary rocks, indicates that atmospheric O_2 attained values during the Permo-Carboniferous as high as 35% O_2. The RROD isotopic modeling assumes that carbon isotope fractionation during photosynthesis globally increases with increasing O_2 level. This agrees with measurements of the carbon isotopic composition of Permo-Carboniferous fossil plants combined with laboratory experiments and calculations on the effect of O_2 on photosynthetic carbon isotope fractionation during plant growth. The principal cause of the high O_2 values was the rise of large vascular land plants that brought about increased O_2 production due to the increased global burial of microbially resistant, lignin-rich organic matter in sediments.

Higher levels of O_2 are consistent with the presence of Permo-Carboniferous giant insects, and preliminary experiments with *Drosophila* indicate that insect body size can increase with elevated O_2. Earlier work purported to show that such high O_2 levels would have led to forest fires so frequent that terrestrial life could not have persisted over geologic time. However, this conclusion is based on burning experiments that consisted only of the ignition of paper strips. Paper is high in cellulose and does not char like natural forest materials. New experimental data, based on fire spreading during the burning of pine wood and pine needles, shows that slightly damp forest materials could resist burning at 35% O_2 and that the degree of moisture is more important to burning than the level of O_2. Nevertheless, elevated O_2 should have led to more frequent and extensive forest fires, and evidence for this is that Permo-Carboniferous plants have morphologies suggestive of adaptation to fire damage.

References

CHAPTER 1

Bergman, N., Lenton, T., and Watson, A., 2003. Coupled Phanerozoic predictions of atmospheric oxygen and carbon dioxide. Geophys. Res. Abstr. Eur. Geophys. Soc. 5:11208.

Berner, E.K., and Berner, R.A., 1996. Global environment: water, air and geochemical cycles. Prentice-Hall, Upper Saddle River, N.J.

Berner, R.A., 1989. Biogeochemical cycles of carbon and sulfur and their effect on atmospheric oxygen over Phanerozoic time. Paleogeogr. Paleoclim. Paleoecol. 75:97–122.

Berner, R.A., 1991. A model for atmospheric CO_2 over Phanerozoic time. Am. J. Sci. 291:339–376.

Berner, R.A., 1994. GEOCARB II. A revised model of atmospheric CO_2 over Phanerozoic time. Am. J. Sci. 294:56–91.

Berner, R.A., 1999. A new look at the long-term carbon cycle. GSA Today 9:1–6.

Berner, R.A., and Kothavala, Z., 2001. GEOCARB III: a revised model of atmospheric CO_2 over Phanerozoic time. Am. J. Sci. 301:182–204.

Berner, R.A., Lasaga, A.C., and Garrels, R.M., 1983. The carbonate-silicate geochemical cycle and its effect on atmospheric carbon dioxide over the past 100 million years. Am. J. Sci. 283:641–683.

Berner, R.A., and Maasch, K.A., 1996. Chemical weathering and controls on atmospheric O_2 and CO_2: fundamental principles were enunciated by J.J. Ebelmen in 1845. Geochim. Cosmochim. Acta 60:1633–1637.

Budyko, M.I., and Ronov, A.B., 1979. Chemical evolution of the atmosphere in the Phanerozoic. Gechem. Intl. 5:643–653.

Crowley, T.J., and Berner, R.A., 2001. CO_2 and climate change. Science 292:870–872.

Ebelmen, J.J., 1845. Sur les produits de la decomposition des especes minérales de la famile des silicates. Annu. Rev. Moines 12:627–654.

Francois, L.M., and Godderis, Y., 1998. Isotopic constraints on the Cenozoic evolution of the carbon cycle. Chem. Geol. 145:177–212.

Garrels, R.M., and Lerman, A., 1984. Coupling of the sedimentary sulfur and carbon cycles—an improved model. Am. J. Sci. 284:989–1007.

Holland, H.D., 1978. The chemistry of the atmosphere and oceans. Wiley, New York.

Holland, H.D., 1984. The chemical evolution of the atmosphere and oceans. Princeton University Press, Princeton, N.J.

IPCC, 2001. Climate change 2001: synthesis report. Intergovernmental Panel on Climate Change, Geneva.

Kashiwagi, H., and Shikazono, N., 2003. Climate change in Cenozoic inferred from carbon cycle model. Geochim. Cosmochim. Acta 66 (suppl. 1):A385.

Kump, L.R., and Arthur, M.A., 1997. Global chemical erosion during the Cenozoic: mass balance constraints on interpretations of Sr isotopic trends. In W.F. Ruddiman, ed., Tectonic uplift and climate change. Plenum Press, New York.

Mackenzie, F.T., Arvidson, R.S., and Guidry, M., 2003. MAGIC: a comprehensive model for earth system geochemical cycling over Phanerozoic time. Geophys. Res. Abst. Eur. Geophys. Soc. 5, no. 13275.

Mackenzie, F.T., and Garrels, R.M., 1966. Chemical mass balance between rivers and oceans. Am. J. Sci. 264:507–525.

Stanley, S.M., 1999. Earth system history. W.H. Freeman, New York.

Sundquist, E.T., 1991. Steady state and non-steady state carbonate-silicate controls on atmospheric CO2. Quaternary Sci. Rev. 10:283–296.

Tajika, E., 1998. Climate change during the last 150 million years: reconstruction from a carbon cycle model. Earth Planetary Sci. Lett. 160:695–707.

Urey, H.C., 1952. The planets: their origin and development: Yale University Press, New Haven, Conn.

Walker, J.C.G., Hays, P.B., and Kasting, J.F., 1981. A negative feedback mechanism for the long-term stabilization of Earth's surface temperature. J. Geophys. Res. 86:9776–9782.

Wallmann, K., 2001. Controls on the Cretaceous and Cenozoic evolution of seawater composition, atmospheric CO_2 and climate. Geochim. Cosmochim. Acta 65:3005–3025.

CHAPTER 2

Aghamiri, R., Schwartzman, D.W., 2002. Weathering rates of bedrock by lichens: a mini watershed study. Chem. Geol. 188(3–4):249–259.

Ague, J.J., 2002. Gradients in fluid composition across metacarbonate layers of the Wepawaug Schist, Connecticut, USA. Contrib. Min. Petrol. 143:38–55.

Algeo, T.J., and Scheckler, S.E., 1998. Terrestrial- marine teleconnections in the Devonian: links between the evolution of land plants, weathering processes, and marine anoxic events. Phil. Trans. R. Soc. Lond. B 353:113–130.

Alt, J.C., and Teagle, D.A.H., 1999. The uptake of carbon during alteration of oceanic crust. Geochim. Cosmochim. Acta 63:1527–1536.

Andrews, J.A., and Schlesinger, W.H., 2001. Soil CO_2 dynamics, acidification and chemical weathering in a temperate forest with experimental CO_2 enrichment. Global Biogeochem. Cycles 15:149–162.

April, R., and Keller, D., 1990. Mineralogy of the rhizosphere in forest soils of the eastern United States. Biogeochemistry 9:1–18.

Arthur, M.A., and Fahey, T.J., 1993. Controls on soil solution chemistry in a subalpine forest in north-central Colorado. Soil Sci. Soc. Am. J. 57:1123–1130.

Augusto, L., Turpault, M.P., and Ranger, J., 2000. Impact of forest tree species on feldspar weathering rates. Geoderma 96:215–237.

Barker, W.W., and Banfield, J.F., 1996. Biologically versus inorganically-mediated weathering reactions: relationships between minerals and extracellular microbial polymers in lithobiontic communities. Chem. Geol. 132:55–69.

Barron, E.J., Fawcett, P.J., Peterson, W.H., Pollard, D., and Thompson, S.L., 1995. A simulation of mid-Creatceous climate. Peloceanography 10:953–962.

Bazzaz, F.A., 1990. The response of natural ecosystems to the rising global CO_2 levels. Annu. Rev. Evol. Syst. 21:167–196.

Berner, E.K., Berner R.A., and Moulton, K.L., 2003. Plants and mineral weathering: present and past. Treatise Geochem. 5:169–188.

Berner, R.A., 1991. A model for atmospheric CO_2 over Phanerozoic time. Am. J. Sci. 291:339–376.

Berner, R.A., 1994. GEOCARB II: a revised model of atmospheric CO_2 over Phanerozoic time. Am. J. Sci. 294:56–91.

Berner, R.A., 1995. Chemical weathering and its effect on atmospheric CO_2 and climate. Rev. Mineral. 31:565–583.

Berner R.A., 1998. The carbon cycle and CO_2 over Phanerozoic time: the role of land plants. Phil. Trans. R. Soc. Lond. B 353:75–82.

Berner, R.A., and Caldeira, K., 1997. The need for mass balance and feedback in the geochemical carbon cycle. Geology 25:955–956.

Berner, R.A., and Cochran, M.F., 1998. Plant-induced weathering of Hawaiian basalts. J. Sed. Res. 68:723–726.

Berner, R.A., and Kothavala, Z., 2001. GEOCARB III: a revised model of atmospheric CO_2 over Phanerozoic time. Am. J. Sci. 301:182–204.

Berner, R.A., Lasaga, A.C., and Garrels, R.M., 1983. The carbonate-silicate geochemical cycle and its effect on atmospheric carbon dioxide and climate, Am. J. Sci. 283:641–683.

Berner, R.A., and Rye, D.M., 1992. Calculation of the Phanerozoic strontium isotope record of the oceans from a carbon cycle model, Am. J. Sci. 292:136–148.

Blum, J.D., Gazis, C.A., Jacobson, A.D., and Chamberlin, C.P., 1998. Carbonate vs. silicate weathering in the Raikhot watershed within the high Himalayan crystalline series. Geology 16:411–414.

Blum, A.E., and Stillings, L.L., 1995. Feldspar dissolution kinetics. Min. Soc. Am. Rev. Mineral 31:291–352.

Bouabid, R., Nater, E.A., and Bloom, P.R., 1995. Characterization of the weathering status of feldspar minerals in sandy soils of Minnesota using SEM and EDX. Geoderma 66:137–149.

Brady, P.V., 1991. The effect of silicate weathering on global temperature and atmospheric CO_2. J. Geophys. Res. 96:18101–18106.

Brady, P.V., Dorn, R.I., Brazel, A.J., Clark, J., Moore, R.B., and Glidewell, T., 1999. Direct measurement of the combined effects of lichen, rainfall, and temperature on silicate weathering. Geochim. Cosmochim. Acta 63:3293–3300.

Brady, P.V., and Gislason, S.R., 1997. Seafloor weathering controls on atmospheric CO_2 and global climate Geochim. Cosmochim. Acta 61:965–973.

Brantley, S.L., and Chen, Y., 1995. Chemical weathering rates of pyroxenes and amphiboles. Min. Soc. Am. Rev. Mineral 31:119–172.

Caldeira, K., 1995. Long term control of atmospheric carbon: low-temperature seafloor alteration or terrestrial silicate-rock weathering? Am. J. Sci. 295:1077–1114.

Chadwick, O.A., Derry, L.A., Vitousek, P.M., Huebert, B.J., and Hedin, L.O., 1999. Changing sources of nutrients during four million years of ecosystem development. Nature 397:491–497.

Chamberlin, T.C., 1899. An attempt to frame a working hypothesis of the cause of glacial periods on an atmospheric basis. J. Geol. 7:545.

Dessert, C., Dupre, B., Francois, L.M., Schott, J., Gaillardet, J., Chakrapani, G., and Bajpai, S., 2001. Erosion of Deccan Traps determined by river geochemistry: impact on the global climate and the Sr-87/Sr-86 ratio of seawater. Earth Planet. Sci. Lett. 188:459–474.

Dessert, C., Dupre, B., Gaillardet, J., Francois, L.M., and Allegre, C.J., 2003. Basalt weathering laws and the impact of basalt weathering on the global carbon cycle. Chem. Geol. 202:257–273.

Drever, J.I., 1994. The effect of land plants on weathering rates of silicate minerals. Geochim. Cosmochim. Acta 58:2325–2332.

Drever, J.I., and Zobrist, J., 1992. Chemical weathering of silicate rocks as a function of elevation in the southern Swiss Alps. Geochim. Cosmochim. Acta 56:3209–3216.

Edmond, J.M., 1992. Himalayan tectonics, weathering processes, and the strontium isotope record in marine limestones. Science 258:1594–1597.

Endal, A.S., and Sofia, S., 1981. Rotation in solar-type stars, I. Evolutionary models for spin-down of the sun. Astrophys. J. 243:625–640.

Fawcett, P.J., and Barron, E.J., 1998. The role of geography and atmospheric CO_2 in long term climate change: results from model simulations for the Late Permian to the Present. In T.J. Crowley and K. Burke, eds., Tectonic boundary conditions for climate reconstructions. Oxford University Press, Oxford.

Fischer, A.G., 1983. The two Phanerozoic subcycles. In W.A. Berggren and J.A. van Couvering, eds., Catastrophies and earth history: the new uniformitarianism. Princeton University Press, Princeton, N.J., pp. 129–150.

Francois, L.M., and Godderis, Y., 1998. Isotopic constraints on the Cenozoic evolution of the carbon cycle. Chem. Geol. 145:177–212.

Gaillardet, J., Dupre, B., Louvat, P., and Allegre, C.J., 1999. Global silicate weathering and CO_2 consumption rates deduced from the chemistry of large rivers. Chem. Geol. 159:3–30.

Galy, A., France-Lanord, C., and Derry, L.A., 1999. The strontium isotopic budget of Himalayan Rivers in Nepal and Bangladesh. Geochim. Cosmochim. Acta 63:1905–1925.

Garrels, R.M., 1967. Genesis of some ground waters from igneous rocks. In: P.H. Abelson, ed., Researches in geochemistry. Wiley, New York, pp. 405–420.

Gensel, P.G., and Edwards, D., 2001. Early land plants and their environments. Columbia University Press, New York.

Gibbs, M.T., Bluth, G.J., Fawcett, P.J., and Kump, L.R., 1999. Global chemical erosion over the past 250 my: variations due to changes in paleogeography, paleoclimate and paleogeology. Am. J. Sci. 299:611–651.

Gibbs, M.T., Rees, P.M., Kutzbach, J.E., Ziegler, A.M., Behling, P.J., and Rowley, D.B., 2002. Simulations of Permian climate and comparisons with climate-sensitive sediments. J. Geol. 110:33–55.

Goldich, S.S., 1938. A study in rock weathering J. Geol. 46:17–58.

Gough, D.O., 1981. Solar interior structure and luminosity variations. Solar Phys. 74: 21–34.

Griffiths, R.P., Baham, J.E., and Caldwell, B.A., 1996. Soil solution chemistry of ectomycorrhizal mats in forest soil. Soil Biol. Biochem. 26:331–337.

Homann, P.S., Van Miegroet, H., Cole, D.W., and Wolfe, G.V., 1992. Cation distribution, cycling, and removal from mineral soil in Douglas fir and red alder forests. Biogeochemistry 16:121–150.

Hovius, N., Strak, C.P., and Allen, P.A., 1997. Sediment flux from a mountain belt derived by landslide mapping. Geology 25:231–234.

Huh, Y., Panteleyev, G., Babich, D., Zaitsev, A., and Edmond, J., 1998. The fluvial geochemistry of the rivers of Eastern Siberia. II. Tributaries of the Lena, Omloy, Yana, Indigirka, Kolyma, and Anadyr draining the collisional/accretionary zone of the Verkhoyansk and Cherskiy ranges. Geochim. Cosmochim. Acta 62:2053–2075.

Hyde, W.T., Crowley, T.J., Tarasov, L., and Peltier, W.R., 1999. The Pangean ice age: studies with a coupled climate-ice sheet model. Climate Dynam. 15:619–629.

Jacobson, A.D., Blum, J.D., Chamberlain, C.P., Craw, D., and Koons, P.O., 2003. Climatic and tectonic controls on chemical weathering in the New Zealand southern Alps. Geochim. Cosmochim. Acta 67:29–46.

Jenny, H., 1941. Factors of soil formation. McGraw-Hill, New York.

Kasting, J.F., and Ackerman, T.P., 1986. Climatic consequences of very high carbon dioxide levels in the Earth's early atmosphere. Science 234:1383–1385.

Kump, L.R., and Arthur, M.A., 1997. Global chemical erosion during the Cenozoic: mass balance constraints on interpretations of Sr isotopic trends. In W.F. Ruddiman, ed., Tectonic uplift and climate change. Plenum Press, New York.

Kutzbach, J.E., and Galimore, R.G., 1989. Pangaean climates: megamonsoons of the megacontinent. J. Geophys. Res. 94:3341–3358.

Lasaga, A.C., 1998. Kinetic theory in the earth sciences. Princeton University Press, Princeton, N.J.

Likens, G.E., Bormann, F.H., Pierce, R.S., Eaton, J.S., and Johnson, N.M., 1977. Biogeochemistry of a forested ecosystem. Springer-Verlag, New York.

Mackenzie, F.T., and Garrels, R.M., 1966. Chemical mass balance between rivers and oceans. Am. J. Sci. 264:507–525.

Meybeck, M., 1987. Global chemical weathering of surficial rocks estimated from river dissolved loads. Am. J. Sci. 287:401–428.

Millot, R., Gaillardet, J., Dupre, B., and Allegre, C.J., 2003. Northern latitude chemical weathering rates: clues from the Mackenzie River Basin, Canada. Geochim. Cosmochim. Acta 67:1305–1329.

Moulton, K.L., West, J., and Berner, R.A., 2000. Solute flux and mineral mass balance approaches to the quantification of plant effects on silicate weathering. Am. J. Sci. 300:539–570.

Otto-Bliesner, B.L., 1993. Tropical mountains and coal formation: a climate model study of the Westphalian (306 Ma). Geophys. Res. Lett. 20:1947–1950.

Otto-Bliesner, B.L., 1995. Continental drift, runoff and weathering feedbacks: implications from climate model experiments. J. Geophys. Res. 100:11537–11548.

Parrish, J.T., Ziegler, A.M., and Scotese, C.R., 1982. Rainfall patterns and the distribution of coals and evaporites in the Mesozoic and Cenozoic. Paleogeogr. Paleoclim. Paleoecol. 40:67–101.

Pavlov, A.A., Hurtgen, M.T., Kasting, J.F., and Arthur, M.A., 2003. Methane-rich Proterozoic atmosphere. Geology 31(1):87–90.

Pavlov, A.A., Kasting, J.F., Brown, L.L., Rages, K.A., and Freedman, R., 2000. Greenhouse warming by CH_4 in the atmosphere of early Earth. J. Geophys. Res.-Planets 105(E5):11981–11990.

Quideau, S.A., Chadwick, O.A., Graham, R.C., and Wood, H.B., 1996. Base cation biogeochemistry and weathering under oak and pine: a controlled long-term experiment. Biogeochemistry 35:377–398.

Raymo, M.E., 1991. Geochemical evidence supporting TC Chamberlin's theory of glaciation. Geology 19:344–347.

Reusch, D.N., and Maasch, K.A., 1998. The transition from arc volcanism to exhumation: weathering of young Ca, Mg, Sr silicates and CO_2 drawdown. In T.J. Crowley and K. Burke, eds., Tectonic boundary conditions for climate reconstructions. Oxford University Press, Oxford.

Richter, F.M., Rowley, D.B., and DePaolo, D.J., 1992. Sr isotope evolution of seawater: the role of tectonics. Earth Planet. Sci. Lett. 109:11–23.

Ronov, A.B., 1993. Stratisfera—Ili Osadochnaya Obolochka Zemli (Kolichestvennoe Issledovanie) (A.A. Yaroshevskii, ed.). Nauka, Moskva.

Ronov, A.B., 1994. Phanerozoic transgressions and regressions on the continents: a quantitative approach based on areas flooded by the sea and areas of marine and continental deposition. Am. J. Sci. 294:802–860.

Schrag, D.P., Berner, R.A., Hoffman, P.F., and Halverson, G.P., 2002. On the initiation of a snowball Earth. Geochem. Geophys. Geosyst. 3:1036.

Scotese, C.R., and Golonka, J., 1995. Wall chart of Phanerozoic paleogeographic reconstructions. Am. Assocn. Petrol. Geologists.

Shaviv, N.J., 2002. Cosmic ray diffusion from the glactic spiral arms, meteorites and a possible climate connection. Phys. Rev. Lett. 89:1–4.

Shukla, J., and Mintz, Y., 1982. Influence of land-surface evapo-transpiration on the Earth's climate. Science 215:1498–1501.

Sleep, N.H., and Zahnle, K., 2001. Carbon dioxide cycling and implications for climate on ancient earth. J. Geophys. Res. 106(E1):1373–1400.

Stallard, R.F., 1995. Relating chemical and physical erosion. Min. Soc. Am. Rev. Mineral 31:543–564.

Stallard, R.F., and Edmond, J.M., 1983. Geochemistry of the Amazon 2: the influence of the geology and weathering environment on the dissolved load. J. Geophys. Res. 88:9671–9688.

Staudigal, H., Hart, S.R., Schmincke, H.-U., and Smith, B.M., 1989. Cretaceous ocean crust at DSDP sites 417 and 418: carbon uptake from weathering vs loss by magmatic outgassing. Geochim. Cosmochim. Acta 53:3091–3094.

Tardy, Y., 1997. Derive des continents paleoclimats et altérations tropicales. Editions BRGM, Orleans, France.

Tardy, Y., N'Kounkou, R., and Probst, J.L., 1989. The global water cycle and continental erosion during Phanerozoic time (570 my). Am. J. Sci. 289:455–483.

Taylor, A.S., and Lasaga, A.C., 1999. The role of basalt weathering in the Sr isotope budget of the oceans Chem. Geol. 161:199–214.

van Breemen, N., Finlay, R., Lundstrom, U., Jongmans, A.G., Giesler, R., and Olsson, M., 2000. Mycorrhizal weathering: a true case of mineral plant nutrition? Biogeochemistry 49:53–67.

Velbel, M.A., 1993. Temperature dependence of silicate weathering in nature: How strong a negative feedback on long term accumulation of atmospheric CO_2 and global greenhouse warming? Geology 21:1059–1062.

Volk, T., 1989. Rise of angiosperms as a factor in long-term climatic cooling. Geology 17:107–110.

Walker, J.C.G., Hays, P.B., and Kasting, J.F., 1981. A negative feedback mechanism for the long term stabilization of Earth's surface temperature. J. Geophys. Res. 86:9776–9782.

Wallmann, K., 2001. Controls on the Cretaceous and Cenozoic evolution of seawater composition, atmospheric CO_2 and climate. Goechim. Cosmochim. Acta 65:3005–3025.

West, A.J., Bickle, M.J., Collins, R., and Brasington, J., 2002. Small-catchment perspective on Himalayan weathering fluxes. Geology 30:355–358.

White, A.F., Blum, A.E., Bullen, T.D., Vivit, D.V., Schulz, M., and Fitzpatrick, J., 1999. The effect of temperature on experimental and natural chemical weathering rates of granitoid rocks. Geochim. Cosmochim. Acta 63:3277–3291.

Willis, K.J., and McElwain, J.C., 2002. The evolution of plants. Oxford University Press, New York.

Wold, C.N., and Hay, W.W., 1990. Estimating ancient sediment fluxes. Am. J. Sci. 290: 1069–1089.

Worsley, T.R., and Kidder, D.L., 1991. 1st-order coupling of paleogeography and CO_2 with global surface temperature and its latitudinal contrast. Geology 19:1161–1164.

CHAPTER 3

Arthur, M.A., Dean, W.E., and Schlanger, S.O., 1985. Variations in the global carbon cycle during the Cretaceous related to climate, volcanism, and changes in atmospheric CO_2. In E.T. Sundquist and W.S. Broecker, eds., The carbon cycle and atmospheric CO_2: natural variations Archean to present. Washington, D.C.: Geophysical Monograph 32, American Geophysical Union, pp. 504–529.

Arthur, M.A., and Sageman, B.B., 1994. Marine black shales-depositional mechanisms and environments of ancient deposits, Annu. Rev. Earth Planet. Sci. 22:499–551.

Arvidson, R.S., and Mackenzie, F.T., 1999. The dolomite problem: control of precipitation kinetics by temperature and saturation state. Am. J. Sci. 299:257–288.

Beerling, D.J., Lake, J.A., Berner, R.A., Taylor, D.W., and Hickey, L.J., 2002. Isotopic evidence for a Permo-Carboniferous high-oxygen event. Geochim. Cosmochim. Acta 66:3757–3767.

Berner, E.K., and Berner, R.A., 1996. The global environment: water, air and geochemical cycles. Prentice-Hall, Upper Saddle River, N.J.

Berner, R.A., 1982. Burial of organic carbon and pyrite sulfur in the modern ocean; its environmental and geochemical significance. Am. J. Sci. 282: 451–473.

Berner, R.A., 1989. Biogeochemical cycles of carbon and sulfur and their effect on atmospheric oxygen over Phanerozoic time. Global Planetary Change 1:97–122.

Berner, R.A., 1994. GEOCARB II: a revised model of atmospheric CO_2 over Phanerozoic time. Am. J. Sci. 294:56–91.

Berner, R.A., 2001. Modeling atmospheric O_2 over Phanerozoic time. Geochim. Cosmochim. Acta 65:685–694.

Berner, R.A., 2003. The long-term carbon cycle, fossil fuels, and atmospheric composition. Nature 426:323–326.

Berner, R.A., and Canfield, D.E., 1989. A model for atmospheric oxygen over Phanerozoic time. Am. J. Sci. 289:333–361.

Berner, R.A., and Kothavala, Z., 2001. GEOCARB III: a revised model of atmospheric CO_2 over Phanerozoic time. Am. J. Sci. 301:182–204.

Berner, R.A., Lasaga, A.C., and Garrels, R.M., 1983. The carbonate-silicate geochemical cycle and its effect on atmospheric carbon dioxide and climate. Am. J. Sci. 283:641–683.

Berner, R.A., and Raiswell, R., 1983. Burial of organic carbon and pyrite sulfur in sediments over Phanerozoic time: a new theory. Geochim. Cosmochim. Acta 47:855–862.

Berry, W.B.N., and Wilde, P., 1978. Progressive ventilation of the oceans—an explanation for the distribution of the lower Paleozoic black shales. Am. J. Sci. 278:257–275.

Bestougeff, M.A., 1980. Summary of world coal resources and reserves. 26th Intl. Geol. Congr. Paris Colloq. c-2 35:353–366.

Blair, N.E., Leithold, E.L., Fort, S.T., Peeler, K.A., Holmes, J.C., and Perkey, D.W., 2003. The persistence of memory: the fate of ancient sedimentary organic carbon in a modern sedimentary system. Geochim. Cosmochim. Acta 67:63–73.

Blum, J.D., Gazis, C.A., Jacobson, A.D., and Chamberlin, C.P., 1998. Carbonate vs silicate weathering in the Raikhot watershed within the high Himalayan crystalline series. Geology 16:411–414.

Bluth, G.J.S., and Kump, L.R., 1991. Phanerozoic paleogeology. Am. J. Sci. 291:284–308.

Boss, S.K., and Wilkinson, B.H., 1991. Planktogenic/eustatic control on cratonic/oceanic carbonate accumulation. J. Geol. 99:497–513.

Caldeira, K., and Berner, R.A., 1999. Seawater pH and atmospheric carbon dioxide. Science 286:2043a.

Canfield, D.E., 1994. Factors influencing organic carbon preservation in marine sediments. Chem. Geol. 114:315–329.

Cerling, T.E., Harris, J.M., MacFadden, B.J., Leakey, M.G., Quade, J., Eisenmann, V., and Ehleringer, J.R., 1997. Global vegetation change through the Miocene/Pliocene boundary. Nature 389:153–158.

Chang, S.B., and Berner R.A., 1999. Coal weathering and the geochemical carbon cycle. Geochim. Cosmochim. Acta 63:3301–3310.

Colman, A.S., and Holland, H.D., 2000. The global diagenetic flux of phosphorus from marine sediments to the oceans: redox sensitivity and the control of atmospheric oxygen levels. In C. Glenn, J. Lucas, and L. Prevot-Lucas, eds., Marine authigenesis: from microbial to global. Soc. Econ. Paleontol. and Mineral. Special Publication 66, pp. 53–75.

Dean, W.E., and Gorham, E., 1998. Magnitude and significance of carbon burial in lakes, reservoirs, and peatlands. Geology 26:535–538.

Drake, J.J., and Wigley, T.M.L., 1975. The effect of climate on the chemistry of carbonate groundwaters. Water Resources Res. 11:958–962.

Eglinton, T.I., Benitez-Nelson, B.C., Pearson, A., McNichol, A.P., Bauer, J.E., and Druffel, E.R.M., 1997. Variability in radiocarbon ages of individual organic compounds from marine sediments. Science 277:796–799.

Emerson, S.R., and Archer, D., 1990. Calcium carbonate preservation in the ocean. Phil. Trans. R. Soc. Lond. A 331:29–40.

Falkowski, P.G., 1997. Evolution of the nitrogen cycle and its influence on the biological sequestration of CO_2 in the ocean. Nature 387:272–275.

Falkowski, P.G., and Raven, J., 1997. Aquatic photosynthesis. Blackwell, Oxford.

Freeman, K.H., and Hayes, J.M., 1992. Fractionation of carbon isotopes by phytoplankton and estimates of ancient CO_2 levels. Global Biogeochem. Cycles 6:185–198.

Gaillardet, J., Dupre, B., Louvat, P., and Allegre, C.J., 1999. Global silicate weathering and CO_2 consumption rates deduced from the chemistry of large rivers. Chem. Geol. 159:3–30.

Guidry, M.W., and Mackenzie, F.T., 2000. Apatite weathering and the Phanerozoic phosphorus cycle. Geology 28:631– 634.

Hansen, K.W., and Wallmann, K., 2003. Cretaceous and Cenozoic evolution of seawater composition, atmospheric O_2 and CO_2. Am. J. Sci. 303:94–148.

Hayes, J.M., Strauss, H., and Kaufman, A.J., 1999. The abundance of ^{13}C in marine organic matter and isotope fractionation in the global biogeochemical cycle of carbon during the past 800 Ma. Chem. Geol. 161:103–125.

Hedges, J.I., Blanchette, R.A., Weliky, K., and Devol, A.H., 1988. Effects of fungal degradation on the CuO oxidation products of lignin: a controlled laboratory study. Geochim. Cosmochim. Acta 52:2717–2726.

Hedges, J.I., Cowie, G.L., Richey, J.E., Quay, P.D., Benner, R., Strom, M., and Forsberg, B.R., 1994. Origins and processing of organic matter in the Amazon River as indicated by carbohydrates and amino acids. Limnol. Oceanogr. 39:743–761.

Hedges, J.I., Hu, F.S., Devol, A.H., Hartnett, H.E., Tsamakis, E., and Keil, R.G.,

1999. Sedimentary organic matter preservation: a test for selective degradation under oxic conditions. Am. J. Sci. 299:529–555.

Hedges, J.I., and Keil, R.G., 1995. Sedimentary organic matter preservation: an assessment and speculative synthesis. Marine Chem. 49:81–115.

Hedges, J.I., Keil, R.G., and Benner, R., 1997. What happens to terrestrial organic matter in the oceans? Organic Geochem. 27:195–212.

Hedges, J.I., and Oades, J.M., 1997. Comparative organic geochemistries of soils and marine sediments. Organic Geochem. 27:319–361.

Holland, H.D., 1978. The chemistry of the atmosphere and oceans. Wiley-Interscience, New York.

Holland, H.D., 1994. The phosphate-oxygen connection. EOS Trans. Am. Geophys. Union 75(3):OS96.

Jenkyns, H.C., 1988. The early Toarcian (Jurassic) anoxic event: stratigraphic, sedimentary, and geochemical evidence. Am. J. Sci. 288:101–151.

Jacobson, A.D., Blum, J.D., Chamberlain, C.P., Craw, D., and Koons, P.O., 2003. Climatic and tectonic controls on chemical weathering in the New Zealand southern Alps. Geochim. Cosmochim. Acta 67:29–46.

Jasper, J.P., and Hayes, J.M., 1990. A carbon isotope record of CO_2 levels during the late Quaternary. Nature 347:462–464.

Lasaga, A.C., and Ohmoto, H., 2002. The oxygen geochemical cycle: dynamics and stability. Geochim. Cosmochim. Acta 66:361–381.

Lenton, T.M., and Watson, A.J., 2000. Redfield revisited 1. Regulation of nitrate, phosphate and oxygen in the ocean. Global Biogeochem. Cycles 14:225–248.

Meybeck, M., 1987. Global chemical weathering of surficial rocks estimated from river dissolved loads. Am. J. Sci. 287:401–428.

Milliman, J.D., 1974. Marine carbonates. Springer-Verlag, New York.

Morse, J.W., and Mackenzie, F.T., 1990. Geochemistry of sedimentary carbonates. Elsevier, Amsterdam.

Pederson, T.F., and Calvert, S.E., 1990. Anoxia vs. productivity: what controls the formation of organic carbon-rich sediments and sedimentary rocks? Bull. Am. Assocn. Petrol. Geol. 74:454–466.

Petsch, S.T., Berner, R.A., and Eglinton, T.I., 2000. A field study of the chemical weathering of ancient sedimentary organic matter. Organic Geochem. 31:475–487.

Petsch, S.T., Eglinton, T.I., and Edwards, K.J., 2001. C-14-dead living biomass: evidence for microbial assimilation of ancient organic carbon during share weathering. Science 292:1127–1131.

Prahl, F.G., Ertel, J.R., Goni, M.A., Sparrow, M.A., and Eversmeyer, B., 1994. Terrestrial organic carbon contributions to sediments on the Washington margin. Geochim. Cosmochim. Acta 58:3035–3048.

Retallack, G.J., 1990. Soils of the past. Unwin-Hyman, London.

Robinson, J.M., 1990. Lignin, land plants and fungi: biological evolution affecting Phanerozoic oxygen balance. Geology 15:607–610.

Robinson, J.M., 1991. Land plants and weathering. Science 252:860.

Ronov, A.B., 1976. Global carbon geochemistry, volcanism, carbonate accumulation and life. Geochem. Intl. (translation of Geokhimiya) 13:172–195.

Ronov, A., Khain, V., and Baukhovsky, 1989. Atlas of lithological-paleogeographical maps of the world: Mesozoic and Cewnozoic of continents. USSR Academy of Science, Leningrad.

Ronov, A.B., Khain, V.E., and Seslavinsky, K.B., 1984. Atlas of lithological-paleogeographical maps of the world: Late Precambrian and Paleozoic of continents. USSR Academy of Science, Leningrad.

Stallard, R.F., 1998. Terrestrial sedimentation and the carbon cycle: coupling weathering and erosion to carbon burial. Global Biogeochem. Cycles 12:231–257.

Stanley, S.M., 1999. Earth system history. W.H. Freeman, New York.

Thomas, L., 2002. Coal Geology. Wiley, New York.

Van Cappellen, P., and Ingall, E.D., 1996. Redox stabilization of the atmosphere and oceans by phosphorus-limited marine productivity. Science 271:493–496.

Veizer, J., Ala, D., Azmy, K., Bruckschen, P., Buhl, D., Bruhn, F., Carden, G.A.F., Diener, A., Ebneth, S., Godderis, Y., Jasper, T., Korte, C., Pawellek, F., Podlaha, O.G., and Strauss, H., 1999. $^{87}Sr/^{86}Sr$, $\delta^{13}C$ and $\delta^{18}O$ evolution of Phanerozoic seawater. Chem. Geol. 161:59–88.

Walker, L.J., Wilkinson, B.H., and Ivany, L.C., 2002. Continental drift and Phanerozoic carbonate accumulation in shallow-shelf and deep-marine settings. J. Geol. 110:75–87.

Wallmann, K., 2001. Controls on the Cretaceous and Cenozoic evolution of seawater composition, atmospheric CO_2 and climate. Goechim. Cosmochim. Acta 65:3005–3025.

Wilde, P., 1987. Model of progressive ventilation of the late Precambrian-early Paleozoic ocean. Am. J. Sci. 287:442–459.

CHAPTER **4**

Barnes, I., Irwin, W.P., and White, D.E., 1978. Global distribution of carbon dioxide discharges and major zones of seismicity. Water Resources Inventory Open-file Report 78-39. U.S. Geological Survey, Washington, D.C.

Beerling, D.J., and Berner, R.A., 2002. Biogeochemical constraints on the Triassic-Jurassic boundary carbon cycle event. Global Biogeochem. Cycles 16(3):1036.

Beerling. D.J., Lomas, M.R., and Grocke, D.R., 2002. On the nature of methane gas-hydrate dissociation during the Toarcian and Aptian oceanic anoxic events. Am. J. Sci 302:28–49.

Berner, R.A., 1971. Principles of chemical sedimentology. McGraw-Hill, New York.

Berner, R.A., 1990. Global CO_2 degassing and the carbon cycle: comment on "Cretaceous ocean crust at DSDP sites 417 and 418: carbon uptake from weathering vs loss by magmatic outgassing" and response to criticism of the BLAG model. Geochim. Cosmochim. Acta 54:2892–2893.

Berner, R.A., 1991. A model for atmospheric CO_2 over Phanerozoic time. Am. J. Sci. 291:339–376.

Berner, R.A., 1994. GEOCARB II. A revised model of atmospheric CO_2 over Phanerozoic time. Am. J. Sci. 294:56–91.

Berner, R.A., and Caldeira, K., 1997. The need for mass balance and feedback in the geochemical carbon cycle. Geology 25:955–956.

Berner, R.A., and Kothavala, Z., 2001. GEOCARB III: a revised model of atmospheric CO_2 over Phanerozoic time. Am. J. Sci. 301:182–204.

Berner, R.A., Lasaga, A.C., and Garrels, R.M., 1983. The carbonate-silicate geochemical cycle and its effect on atmospheric carbon dioxide. Am. J. Sci. 283:641–683.

Bice, K.L., and Marotzke, J., 2002. Could changing ocean circulation have destabilized methane hydrate at the Paleocene/Eocene boundary? Paleoceanography 17:1018.

Bralower, T.J., Premoli-Silva, I., and Malone-Michel, J., 2002. New evidence for abrupt climate change in the Cretaceous and Paleogene; an Ocean Drilling Program expedition to Shatsky Rise, Northwest Pacific. GSA Today 12:4–10.

Brantley, S.L., and Koepenick, K.W., 1995. Measured carbon dioxide emissions from Oldoiyo Lengai and the skewed distribution of passive volcanic fluxes, Geology 23:933–936.

Budyko, M.I., and Ronov, A.B., 1979. Chemical evolution of the atmosphere in the Phanerozoic. Gechem. Intl. 5:643–653.

Caldeira, K., and Rampino, M.R., 1992. Mount Etna CO_2 may affect climate. Nature 355:401–402.

Chaloner, W.G., 1989. Fossil charcoal as an indicator of paleo-atmospheric oxygen level. J. Geol. Soc. Lond. 146:171–174.

Claypool, G., and Kaplan, I.R., 1974. The origin and distribution of methane in marine sediments. In I.R. Kaplan, ed., Natural gases in marine sediments. Plenum Press, New York, pp. 99–139.

Dickens, G.R., 2000. Methane oxidation during the Late Paleocene Thermal Maximum Bull. Soc. Geol. France 171:37–49.

Dickens, G.R., 2001. The potential volume of oceanic methane hydrates with variable external conditions. Org. Geochem. 32:1179–1193.

Dickens, G.R., 2003. A global carbon cycle with gas hydrates and seafloor methane. Geophys. Res. Abstr. 5:04474.

Dickens, G.R., Castillo, M.M., and Walker, J.C.G., 1997. A blast of gas in the latest Paleocene: simulating first-order effects of massive dissociation of oceanic methane hydrate. Geology 25:259–262.

Durand, B., 1980. Kerogen; insoluble organic matter from sedimentary rocks. Editions Technip, Paris.

Engebretson, D.C., Kelley, K.P., Cashman, H.J., and Richards, M.A., 1992. 180 million years of subduction. GSA Today 2:93.

Francois, L.M., Gaillardet, J., and Godderis, Y., 2002. Modelling the Cenozoic evolution of atmospheric CO_2 Geochim. Cosmochim. Acta 66:A243–A243 Suppl.

Francois, L.M., and Godderis, Y., 1998. Isotopic constraints on the Cenozoic evolution of the carbon cycle. Chem. Geol. 145:177–212.

Franck, S., Kossacki, K.J., Von Bloh, W., and Bounama, C., 2002. Long-term evolution of the global carbon cycle: historic minimum of global surface temperature at present. Tellus Ser. B 54:325–343.

Froelich, P.N., Klinkhammer, G.P., Bender, M.L., Luedtke, N.A., Heath, G.R., Cullen, D., Dauphin, D., Hartman, B., and Maynard, V., 1979. Early oxidation of organic matter in pelagic sediments of the eastern equatorial Atlantic: suboxic diagenesis. Geochim. Cosmochim. Acta 43:1075–1090.

Gaffin, S., 1987. Ridge volume dependence of sea floor generation rate and inversion using long term sealevel change. Am. J. Sci. 287:596–611.

Gaína, C., Muller, R.D., and Clark, S., 2003. The evolution of global oceanic crust from Jurassic to present and its contribution to the global carbon budget. Geophys. Res. Abstracts 5:04842.

Gerlach, T.M., 1991. Present-day CO_2 emission from volcanoes. EOS Trans. Am. Geophys. Union 72:249–251.

Hays, J.D., and Pitman, W.C., 1973. Lithospheric plate motion, sea level changes and climatic and ecological consequence. Nature 246:18–22.

Heller, P.L., Anderson, D.L., and Angevine, C.L., 1996. Is the middle Cretaceous pulse of rapid sea-floor spreading real or necessary? Geology 24:491–494.

Heller, P.L., and Angevine, C.L., 1985. Sea level cycles during the growth of Atlantic-type margins. Earth Planet. Sci. Lett. 75:417–426.

Hesselbo, S.P., Grocke, D.R., Jenkyns, H.C., Bjerrum, C.J., Farrimond, P., Bell, H.S.M., and Green, O.R., 2000. Massive dissociation of gas hydrate during a Jurassic oceanic anoxic event. Nature 406:392–395.

Hower, J., Eslinger, E.V., Hower, M.E., and Perry, E.A., 1976. Mechanism of burial metamorphism of argillaceous sediment. I. Mineralogical and chemical evidence. Geol. Soc. Am. Bull. 87:725–737.

Jahren, A.H., Arens, N.C., Sarmiento, G., Guerrero, J., and Amundson, R., 2001. Terrestrial record of methane hydrate dissociation in the Early Cretaceous. Geology 29:159–163.

Kaiho, K., and Saito, S., 1994. Oceanic crust production and climate during the last 100 myr. Terra Nova 6:376–384.

Kerrick, D.M., 2001. Present and past anthropogenic CO_2 degassing from the solid Earth. Rev. Geophys. 39:565–585.

Kerrick, D.M., and Caldeira, K., 1998. Metamorphic CO degassing from orogenic belts. Chem. Geol. 145:213–232.

Kerrick, D.M., and Connolly, J.A.D., 2001. Metamorphic devolatilization of subducted marine sediments and the transport of volatiles into the Earth's mantle. Nature 411:293–296.

Kerrick, D.M., Connolly, J.A.D., and Caldeira, K., 2003. Arc paleo-CO_2 degassing revisited. Eur. Geohys. Soc. Geophys. Res. Abstr. 5:14253.

Kerrick, D.M., McKibben, M.A., Seward, T.M., and Caldeira, K., 1995. Convective hydrothermal CO_2 emission from high heat flow regimes. Chem. Geol. 121:285–293.

Kominz, M.A., 1984. Oceanic ridge volume and sea-level change-an error analysis. In J.S. Schlee, ed., Interregional unconformities and hydrocarbon accumulation. American Association of Petroleum Geologists, Tulsa, Okla. pp. 109–127.

Krull, E.S., and Retallack, G.J., 2000. Delta C-13 depth profiles from paleosols across the Permian-Triassic boundary: evidence for methane release. Geol. Soc. Am. Bull. 112:1459–1472.

Kvenvolden, K.A., 1993. Gas hydrates, geological perspectives and global change. Rev. Geophys. 31:173–187.

Kvenvolden, K.A., 2002. Methane hydrate in the global organic carbon cycle. Terra Nova 14:302–306.

Larson, E.L., 1991. Latest pulse of Earth: evidence for a mid-Cretaceous superplume. Geology 19:547–550.

Martens, C.S., and Berner, R.A., 1977. Interstitial water chemistry of anoxic Long Island Sound sediments 1. Dissolved gases. Limnol. Oceanogr. 22:10–25.

Marty, B., and Tolstikhin, I.N., 1998. CO_2 fluxes from mid-ocean ridges, arcs and plumes. Chem. Geol. 145:233–248.

McLeod, K.G., Smith, R.M.H., Koch, P.L., and Ward, P.D., 2000. Timing of mammal-like reptile extinctions across the Permian-Triassic boundary in South Africa. Geology 28:227–230.

Molina, J.F., and Poli, S., 2000. Carbonate stability and fluid composition in subducted oceanic crust: an experimental study of H_2O-CO_2-bearing basalts. Earth Planet. Sci. Lett. 176:295–310.

Morner, N.-A., and Etiope, G., 2002. Carbon degassing from the lithosphere. Global Planet. Change 33:185–203.

Østergaard, K.K., Anderson, R., Llamedo, M., and Tohidi, B., 2002. Hydrate phase equilibria in porous media: effect of pore size and salinity. Terra Nova 14:307–312.

Parsons, B., 1982. Causes and consequences of the relation between area and age of the ocean floor. J. Geophys. Res. 87:289–302.

Parsons, B., and Sclater, J., 1977. An analysis of the variation of ocean floor bathymetry and heat flow with age. J. Geophys. Res. 82:803–827.

Peltzer, E.T., and Brewer, P.G., 2000. Practical physical chemistry and empirical predictions of methane hydrate stability. In M.D. Cox, ed., Natural gas hydrates in ocean and permafrost environments. Kluwer, Amsterdam, pp. 17–28.

Pitman, W.C. III, 1978. Relationship between eustacy and stratigraphic sequences of passive margins. Geol. Soc. Am. Bull. 89:1389–1403.

Reeburgh, W.S., 1983. Rates of biogeochemical processes in anoxic sediments. Annu. Rev. Earth Planet. Sci. 11:269–298.

Reeburgh, W.S., and Heggie, D.T., 1977. Microbial methane consumption reactions and their effect on methane distributions in freshwater and marine environments. Limnol. Oceanogr. 22:1–9.

Ronov, A.B., 1993. Stratisfera—Ili Osadochnaya Obolochka Zemli (Kolichestvennoe Issledovanie) A.A. Yaroshevskii, ed., Nauka, Moskva.

Rowley, D.B., 2002. Rate of plate creation and destruction: 180 Ma to present. Geol. Soc. Am. Bull. 114:927–933.

Sano, Y., and Williams, S.N., 1996. Fluxes of mantle and subducted carbon along convergent plate boundaries. Geophys. Res. Lett. 23:2749–2752.

Schrag, D.P., 2002. Control of atmospheric CO_2 and climate through earth history. Goldschmidt Conf. Abstr. A688.

Sleep, N.H., and Zahnle, K., 2001. Carbon dioxide cycling and implications for climate on ancient earth. J. Geophys. Res. 106(E1):1373–1400.

Tajika, E., 1998. Climate change during the last 150 million years: reconstruction from a carbon cycle model. Earth Planet. Sci. Lett. 160:695–707.

Tajika, E., and Matsui, T., 1992. Evolution of terrestrial proto-CO_2 atmosphere coupled with thermal history of the Earth. Earth Planet. Sci. Lett. 113:251–266.

Vail, P.R., Mitchum, R.M., and Thompson, S., 1977. Seismic stratigraphy and global changes of sea level, part 4: global cycles of relative changes in sea level. Am. Assocn. Petrol. Geol. Mem. 26:83–98.

Valentine, D.L., 2002. Biogeochemistry and microbial ecology of methane oxidation in anoxic environments: a review. Antoine van Leeuwenhoek Intl. J. Gen. Mole. Biol. 81:271–282.

Valentine, D.L., Blanton, D.C., Reeburgh, W.S., and Kastner, M., 2001. Water column methane oxidation adjacent to an area of active hydrate dissociation, Eel River Basin. Geochim. Cosmochim. Acta 65:2633–2640.

Wallmann, K., 2001. Controls on the Cretaceous and Cenozoic evolution of seawater composition, atmospheric CO_2, and climate. Geochim. Cosmochim. Acta 65:3005–3025.

Welhan, J.A., 1988. Origins of methane in hydrothermal systems. Chem. Geol. 71:183–198.

Welhan, J.A., and Craig, H., 1979. Methane and hydrogen in East Pacific Rise hydrothermal fluids. Geophys. Res. Lett. 6:829–831.

Wilkinson, B.H., and Walker, J.C.G., 1989. Phanerozoic cycling of sedimentary carbonate. Am. J. Sci. 289:525–548.

Wold, C.N., and Hay, W.W., 1990. Estimating ancient sediment fluxes. Am. J. Sci. 290:1069–1089.

Worsley, T.R., Nance, D., and Moody, J.B., 1984. Global tectonics and eustacy for the past 2 billion years. Mar. Geol. 58:373–400.

CHAPTER 5

Andrews, J.E., Tandon, S.K., and Dennis, P.F., 1995. Concentration of carbon dioxide in the Late Cretaceous atmosphere. J. Geol. Soc. Lond. 152:1–3.

Beerling, D.J., 1999. Stomatal density and index: theory and application. In T.P. Jones and N.P. Rowe, eds., Fossil plants and spores: modern techniques. Geological Society, London, pp. 251–256.

Beerling, D.J., and Berner, R.A., 2002. Biogeochemical constraints on the Triassic-Jurassic boundary carbon cycle event. Global Biogeochem. Cycles 16(3):1036.

Beerling, D.J., Lomax, B.H., Royer, D.L., Upchurch, G.R., and Kump, L.R., 2002. An atmospheric pCO_2 reconstruction across the Cretaceous-Tertiary boundary from leaf megafossils. Proc. Natl. Acad Sci. USA 99:7836–7840.

Beerling, D.J., McElwain, J.C., and Osborne, C.P., 1998. Stomatal responses of the living fossil Ginkgo biloba to changes in atmospheric CO_2 concentrations. J. Exp. Botany 49:1603–1607.

Bergman, N., Lenton, T., and Watson, A., 2003. Coupled Phanerozoic predictions of atmospheric oxygen and carbon dioxide. Geophys. Res. Abstr. Eur. Geophys. Soc. 5:11208.

Berner, R.A., 1991. A model for atmospheric CO_2 over Phanerozoic time. Am. J. Sci. 291:339–376.

Berner, R.A., 1994. GEOCARB II. A revised model of atmospheric CO_2 over Phanerozoic time. Am. J. Sci. 294:56–91.

Berner, R.A., 1998. The carbon cycle and CO_2 over Phanerozoic time: the role of land plants: Phil. Trans. R. Soc. B 353:75–82.

Berner, R.A., 2002. Examination of hypotheses for the Permo-Triassic boundary extinction by carbon cycle modeling. Proc. Natl. Acad. Sci. 99:4172–4177.

Berner, R.A., and Kothavala, Z., 2001. GEOCARB III: a revised model of atmospheric CO_2 over Phanerozoic time. Am. J. Sci. 301:182–204.

Berner, R.A., Lasaga, A.C., and Garrels, R.M., 1983. The carbonate-silicate geochemical cycle and its effect on atmospheric carbon dioxide over the past 100 million years. Am. J. Sci. 283:641–683.

Bidigare, R.R., Fluegge, A., Freeman, K.H., Hanson, K.L., Hayes, J.M., Hollander, D., Jasper, J.P., King, L.L., Laws, E.A., Milder, J., Millero, F.J., Pancost, R., Popp, B.N., Steinberg, P.A., and Wakeham, S.G., 1997. Consistent fractionation of ^{13}C in nature and in the laboratory: growth-rate effects in some haptophyte algae. Global Biogeochem. Cycles 11:279–292.

Caldeira, K., and Berner, R.A., 1999. Seawater pH and atmospheric carbon dioxide Science 286:2043a.

Caldeira, K., and Kasting, J.F., 1992. The life span of the biosphere revisited. Nature 360:721–723.

Caldeira, K., and Rampino, M.R., 1990. Carbon dioxide emissions from Deccan volcanism and a K/T boundary greenhouse effect. Geophys. Res. Lett. 17:1299–1302.

Cerling, T.E., 1991. Carbon dioxide in the atmosphere: evidence from Cenozoic and Mesozoic paleosols. Am. J. Sci. 291:377–400.

Cox, J.E., Railsback, L.B., and Gordon, E.A., 2001. Evidence from Catskill pedogenic carbonates for a rapid large Devonian decrease in atmospheric carbon dioxide concentrations. N.E. Geol. Environ. Sci. 23:91–102.

Crowley, T.J., and Berner, R.A., 2001. CO_2 and climate change. Science 292:870–872.

Dessert, C., Dupre, B., Francois, L.M., Schott, J., Gaillardet, J., Chakraparu, G., and Bajpai, 2001. Erosion of Deccan traps determined by river geochemistry: impact on the global climate and the Sr-87/86 ratio of seawater. Earth Planet. Sci. Lett. 188:459–474.

Dickens, G.R., Castillo, M.M., and Walker, J.C.G., 1997. A blast of gas in the latest Paleocene: simulating first-order effects of massive dissociation of oceanic methane hydrate. Geology 25:259–262.

Ekart, D.D., Cerling, T.E., Montanez, I.P., and Tabor, N.J., 1999. A 400 million year carbon isotope record of pedogenic carbonate: implications for paleoatmospheric carbon dioxide. Am. J. Sci. 299:805–827.

Francois, L.M., and Godderis, Y., 1998. Isotopic constraints on the Cenozoic evolution of the carbon cycle. Chem. Geol. 145:177–212.

Freeman, K.H., and Hayes, J.M., 1992. Fractionation of carbon isotopes by phytoplankton and estimates of ancient CO_2 levels. Global Biogeochem. Cycles 6:185–198.

Fry, B., and Wainwright, S.C., 1991. Diatom sources of ^{13}C-rich carbon in marine food webs. Mar. Ecol Prog. Ser. 76:149–157.

Garrels, R.M., and Lerman, A., 1984. Coupling of the sedimentary sulfur and carbon cycles—an improved model. Am. J. Sci. 284:989–1007.

Ghosh, P., Ghosh, P.. and Bhattacharya, S.K., 2001. CO_2 levels in the Late Palaeozoic and Mesozoic atmosphere from soil carbonate and organic matter, Satpura basin, Central India. Palaeogeogr. Palaeoclimatol. Palaeoecol. 170:219–236.

Hansen, K., and Wallmann, K., 2003. Cretaceous and Cenozoic evolution of seawater composition, atmospheric O_2 and CO_2: a model perspective. Am. J. Sci. 303:94–148.

Hayes, J.M., Strauss, H., and Kaufman, A.J., 1999. The abundance of ^{13}C in marine organic matter and isotope fractionation in the global biogeochemical cycle of carbon during the past 800 Ma. Chem. Geol. 161:103–125.

Hemming, N.G., and Hanson, G.N., 1992. Boron isotopic composition and concentration in modern marine carbonates. Geochim. Cosmochim. Acta 56:537–543.

Hinga, K.R., Arthur, M.A., Pilson, M.E.Q., and Whitaker, D., 1994. Carbon isotope fractionation by marine phytoplankton in culture: the effects of CO_2 concentration, pH, temperature, and species. Global Biogeochem. Cycles 8:91–102.

Hollander, D.J., and McKenzie, J.A., 1991. CO_2 control on carbon-isotope fractionation during aqueous photosynthesis: a paleo-CO_2 barometer. Geology 19:929–932.

IPCC, 2001. Climate change 2001: synthesis report. Intergovernmental Panel on Climate Change, Geneva.

Jasper, J.P., and Hayes, J.M., 1990. A carbon isotope record of CO_2 levels during the late Quaternary. Nature 347:462–464.

Kashiwagi, H., and Shikazono, N., 2002. Climate change in Cenozoic inferred from carbon cycle model. Geochim. Cosmochim. Acta 66(suppl 1):A385.

Kump, L.R., 1991. Interpreting carbon-isotope excursions: Strangelove oceans. Geology 19:299–302.

Kump, L.R., and Arthur, M.A., 1997. Global chemical erosion during the Cenozoic: mass balance constraints on interpretations of Sr isotopic trends. In W.F. Ruddiman, ed., Tectonic uplift and climate change. Plenum Press, New York, pp. 203–235.

Kurschner, W.M., 1997. The anatomical diversity of recent and fossil leaves of the durmast oak *Quercus petraea* Liebleinr. *Q. pseudocastanea* Goeppert—implications for their use as biosensors of palaeoatmospheric CO_2 levels. Palaeobot. Palynol. 96:1–30.

Larson, E.L., 1991. Latest pulse of Earth: evidence for a mid-Cretaceous superplume. Geology 19:547–550.

Lee, Y.I., 1999. Stable isotopic composition of calcic paleosols of the Early Cretaceous Hasandong Formation, southeastern Korea. Palaeogeogr. Palaeoclimatol. Palaeoecol. 150:123–133.

Lee, Y.I., and Hisada, K., 1999. Stable isotopic composition of pedogenic carbonates of the Early Cretaceous Shimonoseki Subgroup, western Honshu, Japan. Palaeogeogr. Palaeoclimatol. Palaeoecol. 153:127–138.

Lemarchand, D., Gaillardet, J., Lewin, E., and Allegre, C.J., 2000. The influence of rivers on marine boron isotopes and implications for reconstructing past ocean pH. Nature 408:951–954.

Mackenzie, F.T., Arvidson, R.S., and Guidry, M., 2003. MAGIC: a comprehenive model for earth system geochemical cycling over Phanerozoic time. Geophys. Res. Abstr., Eur. Geophys. Soc. 5:13275.

McElwain, J.C., 1998. Do fossil plants signal palaeoatmospheric CO_2 concentration in the geological past? Phil. Trans. R. Soc. Lond. B 353:83–96.

McElwain, J.C., and Chaloner, W.G., 1995. Stomatal density and index of fossil plants track atmospheric carbon dioxide in the Palaeozoic. Ann. Bot. 76:389–395.

McElwain, J.C., and Chaloner, W.G., 1996. The fossil cuticle as a skeletal record of environmental change. Palaios 11:376–388.

Mora, C.I., and Driese, S.G., 1999. Palaeoenvironment, palaeoclimate and stable carbon isotopes of Palaeozoic red-bed palaeosols, Appalachian Basin, USA and Canada. Spec. Publ. Intl. Assoc. Sedimentol. 27:61–84.

Mora, C.I., Driese, S.G., and Colarusso, L.A., 1996. Middle and Late Paleozoic atmospheric CO_2 levels from soil carbonate and organic matter. Science 271:1105–1107.

Pagani, M., Freeman, K.H., and Arthur, M.A., 1999. Late Miocene atmospheric

CO_2 concentrations and the expansion of C4 grasses. Science 285:876–879.

Palmer, M.R., Pearson, P.N., and Cobb, S.J., 1998. Reconstructing past ocean pH-depth profiles. Science 282:1468–1471.

Pearson, P.N., and Palmer, M.R., 1999. Middle Eocene seawater pH and atmospheric carbon dioxide concentrations. Science 284:1824–1826.

Pearson, P.N., and Palmer, M.R., 2000. Atmospheric carbon dioxide concentrations over the past 60 million years. Nature 406:695–699.

Popp, B.N., Laws, E.A., Bidigare, R.R., Dore, J.E., Hanson, K.L., and Wakeham, S.G., 1998. Effect of phytoplankton cell geometry on carbon isotopic fractionation. Geochim. Cosmochim. Acta 62:69–77.

Rau, G.H., Takahasi, T., Des Marais, D.J., Repeta, D.J., and Martin. J.H., 1992. The relationship between $\delta^{13}C$ of organic matter and $CO_2(aq)$ in ocean surface water: data from a JGOFS site in the northeast Atlantic Ocean and a model. Geochim. Cosmochim. Acta 56:1413–1419.

Retallack, G.J., 1990. Soils of the past. Unwin-Hyman, London.

Rothman, D.H., 2002. Atmospheric carbon dioxide levels for the last 500 million years. Proc. Natl. Acad. Sci. USA 99:4167–4171.

Royer, D.L., 1999. Depth to pedogenic carbonate horizon as a paleoprecipitation indicator? Geology 27:1123–1126.

Royer, D.L., Berner, R.A., and Beerling, D.J., 2001a. Phanerozoic atmospheric CO_2 change: evaluating geochemical and paleobiological approaches. Earth Sci. Rev. 54:349–392.

Royer, D.L., Berner, R.A., Montanez, I.P., Tabor, N.J., and Beerling, D.J., 2004. CO_2 as a primary driver of Phanerozoic climate. GSA Today 14:4–10.

Royer, D.L., Wing, S.L., Beerling, D.J., Jolley, D.W., Koch, P.L., Hickey, L.J., and Berner, R.A., 2001b. Paleobotanical evidence for near present-day levels of atmospheric CO_2 during part of the Tertiary. Science 292:2310–2313.

Salisbury, E.J., 1927. On the causes and ecological significance of stomatal frequency, with special reference to the woodland flora. Phil. Trans. R. Soc. Lond. B 216:1–65.

Sanyal, A., Hemming, N.G., Broecker, W.S., Lea, D.W., Spero, H.J., and Hanson, G.N., 1996. Oceanic pH control on the boron isotopic composition of foraminifera: evidence from culture experiments. Paleoceanography 11:513–517.

Sanyal, A., Hemming, N.G., Hanson, G.N., and Broecker, W.S., 1995. Evidence for a higher pH in the glacial ocean from boron isotopes in foraminifera. Nature 373:234–236.

Schrag, D.P., 2002. Control of atmospheric CO_2 and climate through earth history. Goldschmidt Conf. Abstr. A688.

Sinha, A., and Stott, L.D., 1994. New atmospheric pCO_2 estimates from paleosols during the late Paleocenerearly Eocene global warming interval. Global Planet. Change 9:297–307.

Sundquist, E.T., 1991. Steady state and non-steady state carbonate-silicate controls on atmospheric CO_2. Quat. Sci. Rev. 10:283–296.

Tajika, E., 1998. Climate change during the last 150 million years: reconstruction from a carbon cycle model. Earth Planet. Sci. Lett. 160:695–707.

Twitchett, R.J., Looy, C.V., Morante, R., Visscher, H., and Wignall, P.B., 2001. Rapid and synchronous collapse of marine and terrestrial ecosystems during the end-Permian biotic crisis. Geology 29:351–354.

van der Burgh, J., Visscher, H., Dilcher, D.L., and Kurschner, W.M., 1993. Paleoatmospheric signatures in Neogene fossil leaves. Science 260:1788–1790.

Vengosh, A., Kolodny, Y., Starinsky, A., Chivas, A.R., and McCulloch, M.T., 1991. Coprecipitation and isotopic fractionation of boron in modern biogenic carbonates. Geochim. Cosmochim. Acta 55:2901–2910.

Wallmann, K., 2001. Controls on the Cretaceous and Cenozoic evolution of seawater composition, atmospheric CO_2 and climate. Goechim. Cosmochim. Acta 65:3005–3025.

Woodward, F.I., 1987. Stomatal numbers are sensitive to increases in CO_2 from pre-industrial levels. Nature 327:617–618.

Yapp, C.J., and Poths, H., 1992. Ancient atmospheric CO_2 pressures inferred from natural goethites. Nature 355:342–344.

Yapp, C.J., and Poths, H., 1996. Carbon isotopes in continental weathering environments and variations in ancient atmospheric CO_2 pressure. Earth Planet. Sci. Lett. 137:71–82.

CHAPTER 6

Babrauskas, V., 2002. Ignition handbook: principles and applications to fire, safety engineering, fire investigations, risk management, and forensic science. Fire Science and Technology, Inc., Issaquah, Wash.

Beerling, D.J., and Berner, R.A., 2000. Impact of a Permo-Carboniferous high O_2 event on the terrestrial carbon cycle. Proc. Natl. Acad Sci. USA 97:12428–12432.

Beerling, D.J., Lake, J.A., Berner, R.A., Hickey, L.J., Taylor, D.W., and Royer, D.L., 2002. Carbon isotope evidence implying high O_2/CO_2 ratios in the Permo-Carboniferous atmosphere. Geochim. Cosmochim. Acta 66:3757–3767.

Beerling, D.J., Woodward, F.I., Lomas, M.R., Wills, M.A., Quick, W.P., and Valdes, P.J., 1998. The influence of Carboniferous palaeoatmospheres on plant function: an experimental and modeling assessment. Phil. Trans. R. Soc. Lond. 353:131–140.

Bergman, N., Lenton, T., and Watson, A. 2003. Coupled Phanerozoic predictions of atmospheric oxygen and carbon dioxide. Geophys. Res. Abstr. Eur. Geophys. Soc. 5:11208.

Berner, E.K., and Berner, R.A., 1996. The global environment: water, air and geochemical cycles. Prentice-Hall, Upper Saddle River, N.J.

Berner, R.A., 1987. Models for carbon and sulfur cycles and atmospheric oxygen: application to Paleozoic geologic history. Am. J. Sci. 287:177–190.

Berner, R.A., 2001. Modeling atmospheric O_2 over Phanerozoic time. Geochim. Cosmochim. Acta 65:685–694.

Berner, R.A., Beerling, D.J., Dudley, R., Robinson, J.M., and Wildman, R.A, 2003. Phanerozoic atmospheric oxygen. Annu. Rev. Earth Planet. Sci. 31:105–134.

Berner, R.A., and Canfield, D.E., 1989. A model for atmospheric oxygen over Phanerozoic time. Am. J. Sci. 289:333–361.

Berner, R.A., and Raiswell, R., 1983. Burial of organic carbon and pyrite sulfur in sediments over Phanerozoic time: a new theory. Geochim. Cosmochim. Acta 47:855–862.

Berry, J.A., Troughton, J.H., and Bjorkman, O., 1972. Effect of oxygen concentration during growth on carbon isotope discrimination in C3 and C4 species of Atriplex. Carnegie Inst. Yearbook 71:158–161.

Canfield, D.E., and Teske, A., 1996. Late Proterozoic rise in atmospheric oxygen concentration inferred from phylogenetic and sulphur-isotope studies. Nature 382:127–132.

Carpenter, S.J., and Lohmann, K.C., 1997. Carbon isotope ratios of Phanerozoic marine cements: re-evaluating the global carbon and sulfur systems. Geochim. Cosmochim. Acta 61:4831–4846.

Cerling, T.E., 1991. Carbon dioxide in the atmosphere: evidence from Cenozoic and Mesozoic paleosols. Am. J. Sci. 291:377–400.

Chaloner, W.G., 1989. Fossil charcoal as an indicator of paleoatmospheric oxygen level. J. Geol. Soc. Lond. 146:171–174.

Clark, J.S., Cachier, H., Goldammer, J.G., and Stocks, B., eds., 1997. Sediment records of biomass burning and global change. Springer-Verlag, Berlin.

Cope, M.J., and Chaloner, W.G., 1980. Fossil charcoal as evidence of past atmospheric composition. Nature 283:647–649.

Deeming. J., Burgan, R., and Cohen, J.D., 1977. National fire danger rating system—1978. USDA General Technical Report INT-39. USDA Forest Service, Ogden, Ut.

Dudley, R., 1998. Atmospheric oxygen, giant Paleozoic insects and the evolution of aerial locomotor performance. J. Exp. Biol. 201:1043–1050.

Dudley, R., 2000. The biomechanics of insect flight: form, function, evolution. Princeton University Press, Princeton, N.J.

Ebelmen, J.J., 1845. Sur les produits de la decomposition des especes minérales de la famile des silicates. Annu. Rev. Moines 12:627–654.

Edmond, J.M., Measures, C., McDuff, R.E., Chan, L.H., Collier, R., et al., 1979. Ridge crest hydrothermal activity and the balances of the major and minor elements in the ocean: the Galapagos data. Earth. Planet. Sci. Lett. 46:1–18

Farquhar, G.D., and Wong, S.C., 1984. An empirical model of stomatal conductance. Aust. J. Plant Physiol. 11:191–210.

Francois, L.M., and Gerard, J.C., 1986. A numerical model of the evolution of ocean sulfate and sedimentary sulfur during the past 800 million years. Geochim. Cosmochim. Acta 50:2289–2302.

Frazier, M.R., Woods, H.A., and Harrison, J.F., 2001. Interactive effects of rearing temperature and oxygen on the development of *Drosophila melanogaster*. Physiol. Biochem. Zool. 74:641–650.

Fridovich, I., 1975. Oxygen: boon or bane. Am. Sci. 63:54–59.

Garrels, R.M., and Lerman, A., 1984. Coupling of the sedimentary sulfur and carbon cycles—an improved model. Am. J. Sci. 284:989–1007.

Garrels, R.M., Lerman, A., and Mackenzie, F.T., 1976. Controls of atmospheric O_2 and CO_2-past, present, and future. Am. Sci. 64:306–315.

Garrels, R.M., and Perry, E.A., 1974. Cycling of carbon, sulfur and oxygen through geologic time. In E. Goldberg, ed., The sea, vol. 5. Wiley, New York, pp. 303–316.

Graham, J.B., Dudley, R., Aguilar, N., and Gans, C., 1995. Implications of the late Palaeozoic oxygen pulse for physiology and evolution. Nature 375:117–120.

Hansen, K.W., and Wallmann, K., 2003. Cretaceous and Cenozoic evolution of seawater composition, atmospheric O_2 and CO_2. Am. J. Sci. 303:94–148.

Harrison, J.F., and Lighton, J.R.B., 1998. Oxygen-sensitive flight metabolism in the dragonfly *Erythemis simplicicollis*. J. Exp. Biol. 201:1739–1744.

Hayes, J.M., Strauss, H., and Kaufman, A.J., 1999. The abundance of ^{13}C in marine organic matter and isotope fractionation in the global biogeochemical cycle of carbon during the past 800 Ma. Chem. Geol. 161:103–125.

Holland, H.D., 1978. The chemistry of the atmosphere and oceans. Wiley, New York.

Jones, T.P., and Chaloner, W.G., 1991. Fossil charcoal, its recognition and palaeoatmospheric significance. Palaeogeogr. Palaeoclimatol. Palaeoecol. 97:39–50.

Kadko, D., 1996. Radioisotopic studies of submarine hydrothermal vents. Rev. Geophys. 34:349–366.

Komrek, E.V., 1973. Ancient fires. In Proceedings of the Annual Tall Timbers Fire Ecology Conference, June 8–9, Lubbock, Texas. Tall Timbers Research Station, Tallahassee, Fla., pp. 219–240.

Kump, L.R., 1988. Terrestrial feedback in atmospheric oxygen regulation by fire and phosphorus. Nature 335:152–154.

Kump. L.R., 1989. Alternative modeling approaches to the geochemical cycles of carbon, sulfur and strontium isotopes. Am. J. Sci. 289:390–410.

Kump, L.R., and Garrels, R.M., 1986. Modeling atmospheric O_2 in the global sedimentary redox cycle. Am. J. Sci. 286:336–360.

Lane, N., 2002. Oxygen: the molecule that made the world. Oxford University Press, Oxford.

Lasaga, A.C., 1989. A new approach to the isotopic modeling of the variation of atmospheric oxygen through the Phanerozoic. Am. J. Sci. 298:411–413.

Lenton, T.M., 2001. The role of plants, phosphorus weathering and fire in the rise and regulation of atmospheric oxygen. Global Change Biol. 7:613–629.

Mackenzie, F.T., Arvidson, R.S., and Guidry, M., 2003. MAGIC: a comprehenive model for earth system geochemical cycling over Phanerozoic time. Geophys. Res. Abstr. Eur. Geophys. Soc. 5:13275.

McAlester, A.L., 1970. Animal extinctions, oxygen consumption and atmospheric history. J. Paleontology 44:405–409.

Morse, J.W., Millero, F.J., Cornwell, J.C., and Rickard, D., 1987. The chemistry of the hydrogen sulfide and iron sulfide systems in natural waters. Earth Sci. Rev. 24:1–42.

Morton, J.L., and Sleep, N.H., 1985. A mid-ocean ridge thermal model-constraints on the volume of axial hydrothermal heat-flux. J. Geophys. Res. Solids 90:134–153.

Nelson, M., 2001. A dynamical systems model of the limiting oxygen index test: II. Retardancy due to char formation and addition of inert fillers. Combust. Theory Model. 5:59–83.

Peters-Kottig, W., Strauss, H., and Kerp, H., 2003. Land plants, carbon isotopes and the late Paleozoic carbon cycle. Geophys. Res. Abstr. Eur. Geophys. Soc. 5:10794.

Petsch, S.T., 1999. Comment on "Carbon isotope ratios of Phanerozoic marine cements: re-evaluating global carbon and sulphur systems," [S.J. Carpenter and K.C. Lohmann (1997)]. Geochim. Cosmochim. Acta 61:307–310.

Petsch, S.T., and Berner, R.A., 1998. Coupling the long-term geochemical cycles of carbon, phosphorus, sulfur, and iron: the effect on atmospheric O_2 and the isotopic records of carbon and sulfur. Am. J. Sci. 298:246–262.

Rashbash, D., and Langford, B., 1968. Burning of wood in atmospheres of reduced oxygen concentration. Combust. Flame 12:33–40.

Robinson, J.M., 1989. Phanerozoic O_2 variation, fire, and terrestrial ecology. Palaeogeogr. Palaeoclimatol. Palaeoecol. 75:223–240.

Robinson, J.M., 1991. Phanerozoic atmospheric reconstructions: a terrestrial perspective. Global Planet. Change 5:51–62.

Ronov, A.B., 1976. Global carbon geochemistry, volcanism, carbonate accumulation and life. Geochem. Intl. 13:172–195.

Rowley, D.B., 2002. Rate of plate creation and destruction:180 Ma to present. Geol Soc. Am. Bull. 114:927–933.

Scott, A.C., 2000. The Pre-Quaternary history of fire. Palaegeogr. Palaeoclimatol. Palaeoecol. 164:335–371.

Strauss, H., 1999. Geological evolution from isotope proxy signals—sulfur. Chem. Geol. 161:89–101.

Tewarson, A., 2000. Nonmetallic material flammability in oxygen enriched atmospheres. J. Fire Sci. 18:183–214.

Veizer, J., Ala, D., Azmy, K., Bruckschen, P., Buhl, D., Bruhn, F., Carden, G.A.F., Diener, A., Ebneth, S., Godderis, Y., Jasper, T., Korte, C., Pawellek, F., Podlaha, O.G., and Strauss, H., 1999. $^{87}Sr/^{86}Sr$, $\delta^{13}C$ and $\delta^{18}O$ evolution of Phanerozoic seawater. Chem. Geol. 161:59–88.

Veizer, J., Holser, W.T., and Wilgus, C.K., 1980. Correlation of C-13-C-12 and S-34-S-32 secular variations. Geochim. Cosmochim. Acta 44:579–587.

Vogel, S., 1994. Life in moving fluids: the physical biology of flow. Princeton University Press, Princeton, N.J.

Walker, J.C.G., 1986. Global geochemical cycles of atmospheric oxygen. Mar. Geol. 70:159–174.

Watson, A.J., 1978. Consequences for the biosphere of grassland and forest fires. PhD dissertation, Reading University, Reading, UK.

Watson, A., Lovelock, J.E., and Margulis, L., 1978. Methanogenesis, fires, and the regulation of atmospheric oxygen. Biosystems 10:293–298.

Wildman, R.A., Hickey, L.J., Dickinson, M.B., Berner, R.A., Robinson, J.M., Dietrich, M., Essenhigh, R.H., and Wildman, C.B., 2004. Burning of forest materials under late Paleozoic high atmospheric oxygen levels. Geology (in press).

Index

150   Index

stomatal density. *See* stomatal index
stomatal index
    as CO_2 indicator, 92–94
    definition of, 92
strontium isotopes and weathering,
    14–16
sulfur in rocks, 104–106
sulfur cycle
    modeling of, 106–107
    principal reactions of, 101–103
sulfur isotopes and O_2 modeling,
    106–107
surficial system, 9, 10, 72

volcanic rocks
    abundance over time of, 61–63
    areal density of, 60–61
    as indicators of degassing, 61–63

weathering
    of Ca and Mg silicates, 6, 13–39
    of carbonates, 7, 8, 52–54
    as control on CO_2, 6, 26
    effect of angiosperms on, 23–24
    effect of carbon dioxide on, 25–
        31
    effect of land area on, 36
    effect of lithology on, 36–37
    effect of plants on, 18–25
    effect of temperature on, 25–31
    in Hawaiian Islands, 22
    in Iceland, 22
    by lichens and bryophytes, 21
    and mountain uplift, 13–18
    of organic matter, 50–52
    and physical erosion, 13–18
    and strontium isotopes, 14–16